Borrowed Earth, Borrowed Time

Healing America's Chemical Wounds

Borrowed Earth, Borrowed Time

Healing America's Chemical Wounds

Glenn E. Schweitzer

Plenum Press • New York and London

Library of Congress Cataloging-in-Publication Data

Schweitzer, Glenn E., 1930-
Borrowed earth, borrowed time : healing America's chemical wounds / Glenn E. Schweitzer.
p. cm.
Includes bibliographical references and index.
ISBN 0-306-43766-X
1. Pollution--Environmental aspects--United States. 2. Environmental protection--United States. 3. Industry--Environmental aspects--United States. I. Title.
TD180.S39 1991
363.73'0973--dc20 90-29054
CIP

ISBN 0-306-43766-X

Plenum Press is a division of Plenum Publishing Corporation
233 Spring Street, New York, N.Y. 10013

Printed in the United States of America

Preface

The rules of the game are what we call Nature. The player on the other side is hidden from us. We know that his play is always fair, just, and patient. But also we know, to our cost, that he never overlooks a mistake, or makes the smallest allowance for ignorance.
—Thomas Huxley

High ozone levels in the largest American cities show little sign of abatement. Chemicals from waste dumps in some metropolitan areas are draining into backyards. Pesticides are leaching through the soil into drinking water wells in our agricultural belts. Medical wastes are fouling beaches on the Eastern seaboard. These and other toxic hazards are jeopardizing life-styles and even human lives throughout the country.

Dying forests in the mountains above Los Angeles have become victims of urban chemicals. The acidic lakes of New England accumulate the fallout from industrial facilities hundreds of miles away. Damaged wildlife refuges in Nebraska and California testify to the harmful side effects of improper agricultural practices. The nation's natural resource base is being slowly eroded with no end in sight.

Refrigerators were designed to cool, but their coolants seep into the atmosphere and thin the Earth's protective shield. Power plants sustain all modern industrial societies, but their emissions may lead to global warming and trigger flooding of coastal areas. Giant supertankers provide nations easy access to oil, our black gold, but marine accidents and leaking liquids from ocean traffic pollute the coastal areas of the world. The chemical by-products of many of society's proudest technological achievements are degrading distant corners of the planet.[1]

During the past decade, so many health and ecological alarms have been sounded that historic apathy over the state of the environment has turned into widespread anxiety. All politicians now espouse strong environmental platforms. States and communities are searching for cleaner industries. And every industrialized nation currently embraces the priority of environmental protection while promising aggressive regulation in the future.

Americans want clean air, yet few are enthusiastic about the inconvenience of automobile exhaust inspections. Everyone strongly supports clean water, although no one wants to live near a sewage treatment plant. Garbage and chemical wastes must be placed out of sight, but not near my property, each of us proclaims.

Relatively few households take the trouble to reduce or even segregate the hundreds of pounds of refuse we each generate every year or to limit wasteful uses of energy. We can't find time to participate in public discussions of environmental preservation. Everybody resists paying higher taxes to clean up the environment. We simply will not sacrifice any aspect of our personal lives as the price for putting a lid on further contamination.

It is easy for Americans to point to big business as the source of environmental problems. Also, they applaud the press and the Congress when these institutions place the responsibility for environmental tragedies at the doorstep of the United States Environmental Protection Agency (EPA). Those who suffer personal damage from environmental insults increasingly seek compensation through the courts from these same "responsible" parties—big business and the government. Business and the government certainly cannot dodge their responsibilities but neither can individual citizens.

*   *   *

Important questions that each of us must face are: How serious are the environmental problems engulfing the nation? Are we dooming future generations to a dangerous life in a cul-de-sac of intolerable contamination? Can we reverse the current practices that are clouding our water, browning the air, and laying waste to the land? Can science and technology come to the rescue? Or are our children and grandchildren destined to drink bottled water, swelter under a greenhouse blanket, and live among a patchwork of toxic cemeteries which are fenced off from everyday life? Are we indeed approaching an environmental apocalypse?

Perhaps the significance of a few toxic incidents has been blown out of proportion. Aren't medical experts who warn us of the harmful effects of modern chemicals trained to be cautious? If forests and farmlands have been able to withstand natural fires and floods for centuries, they surely must have the resilience to recover from man-made stresses.

For those with polluted wells, dead orchards, and contaminated waterfronts, the environmental problems are all too real. However, to the vast majority of Americans who have yet to be personally harmed by environmental contaminants, the constant media reports of toxic disasters may seem like a hypnotic obsession. Still, even they cannot help but to relate very directly to banner headlines like "Hypersensitivity to Chemicals Called Rising Health Problem; Some Cannot Adapt to Low Doses of Toxics, Study Says."[2] We are far less likely to remember reports of the ever-increasing life expectancy of almost every modern society or to take pride in the great progress of our nation in cleaning up dirty rivers and in reducing air pollution during the past two decades. Whether crisis or obsession, toxic mists are around us—mentally and physically. They can no longer be ignored by anyone.

* * *

First of all, we need to accept a new perspective toward nature. For years we have been hearing about a "fragile" planet which sustains life. The planet is not fragile. Man is fragile.

As noted in *Newsweek,* ". . . 99 percent of the creatures ever to come into existence have vanished. Nature doesn't care if the globe is populated by trilobites or thunder lizards or people or six-eyed telepathic slugs . . . . Should man sour the environmental conditions now slanted in our favor, creatures will rise up in our stead that thrive on murky greenhouse air or dine on compounds human metabolisms find toxic."[3]

We are not masters of the land and water. We belong to the Earth as much as the Earth belongs to us. The next century will be a decisive period in determining just how much longer our species will be supported by this planet.

Economic development that regenerates rather than erodes nature must replace the traditional emphasis on near-term growth. We need sustainable development, not only for the benefit of the planet but also for the survival of mankind. Such an approach can belie the often quoted doomsday warning of Albert Schweitzer: "Man has lost the capacity to foresee and forestall. He will end by destroying the earth."

* * *

At the beginning of the modern environmental movement in the early 1970s, ecologists persuasively argued that the only sensible approach to environmental protection was a holistic approach. In brief, the biosphere of the planet which encompasses both nature and mankind must be viewed as an integrated system wherein each action by humans causes reactions not only from other humans but also from nature. In the words of Indian Chief Seattle, "Man did not weave the web of life. He is merely a strand in it. Whatever he does to the web, he does to himself."

However, understanding the interrelationships between human beings and nature is a formidable task. Indeed, understanding the relationships between human beings themselves is often impossible.

In the United States, the approach to environmental protection has been to talk about a holistic environment but to take actions which address more easily defined problems. Of necessity, we have compartmentalized complicated ecological and human systems into manageable components—land use planning, agricultural policies, air pollution control, and preservation of wetlands, for example. In each of these areas, government agencies and interest groups at the national and local levels make different types of trade-offs.

Not surprisingly, our nation now has many fragmented and sometimes inconsistent environmental policies and regulations which a variety of entrenched economic interests constantly challenge. Individually, the policies respond to different sets of problems within different political and economic constraints. Collectively they will, we hope, limit the most worrisome types of environmental pollution without unnecessarily inhibiting the aspirations of all citizens for more prosperous lives.

The appropriateness and effectiveness of these disparate approaches to preserving the environment—approaches which in effect call for "social engineering" of life-styles and economic priorities—are central themes of this book. How can the government adjust its policies during the 1990s to best ensure the future compatibility of society's activities and the capacity of the planet to support us—locally, nationally, and internationally?

* * *

Noxious chemicals resulting from human activities have been repeatedly indicted as the root of many environmental problems—DDT, mercury, PCBs, dioxin, and the list goes on. Some chemicals cause cancer and other harmful, even deadly, diseases. Several chemicals cause acid rain, and still others cause global warming.

But life is based on chemicals. They are responsible for our current standard of living. They undergird agricultural productivity. They provide the weapons for combatting disease. They produce the energy which enables industrial societies to function. Almost every significant consumer product depends on man-made chemicals. In short, without chemicals we would have to return to the days prior to the industrial revolution.

This book is about chemicals—about their use and their abuse. It is about our nation's many efforts to control chemicals, those that have useful properties and those that are simply unwanted wastes. Many of these chemicals also occur in nature. Others do not. Indeed, some chemicals which have always been present in nature are more threatening as food contaminants, for example, than chemicals produced by industry. Yet we seldom question the wholesomeness or safety of natural food products.

Several million chemicals have been discovered in the laboratories of the world. About 70,000 are produced commercially, and many more are found in nature. At very high concentration, almost every chemical is harmful—even the chemical ingredients of salt, sugar, or sand. Perhaps it is true, as industry leaders often contend, that the government is trying too hard to control too many chemicals which are simply uncontrollable or which do not need to be controlled. Still, the government can certainly take constructive steps to limit damage from some of the most harmful substances.

Radioactive chemicals are of special concern to many Americans. However, this topic is too large to be considered in detail in this book.

Nevertheless, energy production is central to many environmental concerns, and the role of nuclear power in the future energy mix of the United States and the world cannot be ignored in this discussion. Also, a brief examination of the handling of nuclear wastes together with a review of the disposal of chemical wastes which are not radioactive may help bridge the gulf between nuclear and chemical specialists. In this regard, only recently have environmental experts recognized the extensive comingling of nuclear wastes with nonnuclear, but hazardous, wastes. This comingling exists at many of the large waste disposal sites throughout the country.

With further regard to the scope of this book, conservation measures to perpetuate forests, to preserve parks and wilderness areas, to sustain soils, and to protect endangered plants and animals are fundamental to the protection of the environment and are increasingly entwined with pollution issues. However, conservation is also a very broad topic and deserves extensive treatment on its own. The controversies over toxic chemicals in the environment are sufficiently far-reaching to warrant limiting this book to the chemical aspects of environmental protection.

* * *

This book reviews efforts to control hazardous chemicals from the early 1970s into the very beginning of the 1990s. Many of the issues which were hotly debated 15 years earlier are still on the national agenda and are surrounded with more controversy than ever before. Recalling some of the earlier debates should help clarify the choices currently before the American people and the world community.

The emphasis is on environmental goals and on general policies and approaches in search of these goals. "Unless we know where we are going, we will never arrive at our destination" is a time-worn saying which is particularly appropriate for environmental protection.

At the same time, environmental issues teem with controversy over details—in the fine print of the laws, in the *Federal Register* announcements of new regulations, in enforcement orders, and in court rulings. In the environmental field the gaps are enormous between the politicians, the experts, and the public in understanding the facts, in recognizing the uncertainties, in developing a basis for reaching decisions laden with value judgments, and in translating these judgments into implementation activities.

However, the goals must be clear even to identify the important facts and uncertainties themselves and to frame the judgments which must be made. If a consensus cannot be reached on the appropriate goals and objectives, then the nation is in serious trouble. The importance of such a consensus is underscored by one industrial estimate that each American family already pays $150 per month in direct and hidden environmental costs, expenditures which we can ill-afford to misdirect.

* * *

This book draws extensively on my experiences as the first Director of the EPA's Office of Toxic Substances from 1973 to 1977 and as the Director of the EPA's Environmental Monitoring Systems Laboratory in Las Vegas from 1980 to 1985. During these periods I profited greatly from the observations and suggestions of many EPA colleagues in Washington, in Las Vegas, and in the EPA regional offices and laboratories throughout the country.

While serving in the EPA, I spent most of my time on the front lines of environmental protection—assessing chemical spills and waste disposal practices in the field, developing assessment hardware in the laboratory, persuading industry to restrain chemical discharges at plants from New England to Southern California, participating in meetings with irate citizens from New York to Texas, fending off aggressive journalists, and reaching compromises with congressional committees and environmental advocacy groups. I felt comfortable at that time that we knew what we were doing, because we had good staffs and extensive financial resources to tap the leading experts throughout the country.

In retrospect, I feel less comfortable today. We underestimated the gaps in our knowledge and the difficulty in comprehending the significance of environmental trends. Also, we simply did not look ahead for more than two or three years. This book now provides an opportunity to take that longer-term view.

Of the many hundreds of documents I reviewed during my preparation of the manuscript, one source of information has been particularly useful, namely, the American Chemical Society. It is the world's largest professional society with over 140,000 members drawn from universities, government, and industry. The society's many excellent publications provided me with new perspectives and insights into the challenges, opportunities, and pitfalls in regulating chemical activities and pointed the way to relevant publications of other organizations as well.

While drawing on the wisdom and perceptions of many others, I alone am responsible for any shortcomings in this unvarnished record of developments in the environmental field which I have witnessed over the years. A few friends commented on an early draft of the book, and I am particularly grateful for their help. Also, Linda Regan of Plenum Publishing Corporation provided a thorough, unrestrained critique of that draft, and her insightful suggestions added immeasurably to the logic and readability of the text.

Included in the discussions are many personal experiences which I hope will be accepted in the spirit in which they are offered. I am but one of hundreds of thousands of Americans who have been vitally concerned with protecting the environment. I have been very fortunate to have had some unique experiences, and this book provides an opportunity to share with a broader audience a few personal observations that seem to have special relevance to controlling chemicals in the years ahead.

Contents

1. Toxic Chemicals Move to Center Stage *1*

The Environmental Awakening of a Nation *1*
The Environmental Movement Restores the American Spirit *4*
From Eyesores and Odors to Toxics and Cancer *7*
Searching for Legislative Rationality: The Toxic Substances Control Act *10*
A Passive President Accepts the New Toxics Legislation *16*
The Education of the Revolving-Door Regulators *18*
Impatience Leads to Alternative Routes for Controlling Toxics *23*
Ecological Concerns Return to the Top of the Regulatory Agenda *28*
The Many Dimensions of the Toxics Problem *30*
Controlling Chemical Pollution in the 1990s *33*

2. Controlling 3000 Chemicals: One by One *35*

The Toxic Chemical of the Month: Hexachlorobenzene *35*
Containing One of the Most Toxic Pollutants: Dioxin *40*
Setting National Standards for Toxic Air Pollutants *44*
Regulating Categories of Chemicals *48*
Waste Products Laden with Hazardous Chemicals *52*
150 Chemicals Told the Story of Love Canal *55*
Controlling the Most Toxic Chemicals with Concern for Many More *61*

3. The Uncertainty of Risk but the Reality of Cost *65*

Toxic Trouble in the Plastics Industry: Vinyl Chloride *65*
A Threat to the Health of Newborn Infants: PCBs *72*
Quantifying the Risks from Toxic Chemicals *76*
On Being Exposed to Mixtures of Chemicals *84*
Chemical Threats to Ecological Resources *86*
Documenting Uncertainty Leads to Informed Decisions *89*
Science, Values, and Environmental Priorities *91*
The Judiciary Speaks Out on Risk and Uncertainty *94*
Reducing Risks during the 1990s *98*

4. Explaining Risks to an Aroused Public *103*

The Power of Television *103*
The Continuing Impact of the Press *107*
Seeking Consensus but Encountering Controversy at Public Meetings *109*
The Conventional Wisdom for Communicating with the Public *115*
Withholding Scientific Data from the Public *118*
Mediation Tempers Confrontation *122*
Conducting the Business of Government in a Glass House *125*

5. Cleaning Up the Wastes of an Industrial Economy *129*

Toxic Wastes Invade Every City and County *129*
Expectations and Disappointments of Superfund *135*
Safe Disposal of Hazardous Wastes *143*
Two Hundred Thousand Leaking Tanks *147*
The Fears and the Reality of Nuclear Waste *149*
Environmental Neglect at Nuclear Weapons Plants *156*
Protecting Our Groundwater Resources *159*

6. The States Begin to Take Charge *165*

A Soft Collar Protects the Chesapeake Bay *165*

Setting the Environmental Agenda and Paying the Bill *168*
Recognizing the Impacts of Chemical Runoff *173*
States Want to Get Tough on Toxics *176*
Keeping Pesticides Out of Groundwater *180*
The Importance of an Informed Public *182*
The States Prepare for the Long Haul *185*

7. The Greening of Industry ***189***

A Large Company Looks Ahead but Stumbles with the Present *189*
Environmental Outrage Engulfs a Small Company *193*
The Environmental Consciousness of the Chemical Industry *196*
Industry Efforts to Reduce Toxic Wastes *200*
Industry Reaches Out to the Public *203*
Industry Braces for Transportation Accidents *206*
The Changed Character of American Industry *208*

8. Exceeding the Absorptive Limits of the Global Commons ***213***

Erosion of the Earth's Protective Shield *213*
Gaseous Pollutants Warm Up the Earth *221*
Coping with Fossil Fuels *227*
Reviving the Nuclear Option *229*
Too Many People around the World *233*
The Growing Role of Eco-Diplomacy *236*
Environmental Negotiations at All Levels *240*
International Ecological Security *244*

9. The Power and Limitations of Science and Technology ***249***

Lasers Light Up the Groundwater *249*
Finding Pollutants from Airplanes and Satellites *253*
Technology and the Greenhouse Gases *257*
Farmers Seek Environmental Acceptance *261*

The Promise of Biotechnology *264*
Technological Opportunities and Social Engineering *267*

10. Living with Toxic Risks ***271***

End Notes ***281***

Index ***289***

Borrowed Earth, Borrowed Time

Healing America's Chemical Wounds

1 Toxic Chemicals Move to Center Stage

Chemistry: a science that deals with the composition, structure, and properties of substances and with the transformations that they undergo.
—Webster's Ninth New Collegiate Dictionary

Any substance can be toxic when administered in sufficient amount. Even essential nutrients like vitamins are poisons when taken in too large a quantity. But this does not mean that the amounts of substances to which an organism normally may be exposed pose a significant hazard or, indeed, any hazard at all.
—American Chemical Society

The Congress finds that (1) human beings and the environment are being exposed each year to a large number of chemical substances and mixtures; (2) among the many chemical substances and mixtures . . . are some whose manufacture, processing, distribution in commerce, use, or disposal may present an unreasonable risk of injury to health or the environment. . . .
—Toxic Substances Control Act of 1976

The Environmental Awakening of a Nation

The conference room in the new Executive Office Building across Pennsylvania Avenue from the White House was jammed. I managed to find standing room at a considerable distance from the White House staffers who were beaming with a sense of personal accomplishment at the far end of the table. They had called together representatives of about 20 government agencies to inform them of the procedures for implementing the newly enacted National Environmental Policy Act of

1969 which the presidential staff had shepherded through the Congress.

The White House aides emphasized the importance of Environmental Impact Statements. The law required that such statements be prepared by government agencies prior to any federal action that could have significant environmental side effects. They announced that the White House was assembling a new staff for reviewing these statements and the related plans to prevent harmful effects before the agencies could carry out their proposed projects.

At the meeting I was representing the Agency for International Development (AID), the foreign assistance arm of the United States government. I had clear instructions from the Agency: keep these instant environmentalists out of the foreign aid business. The whole purpose of foreign assistance, we would argue, was to improve human environments in distant lands. Environmental Impact Statements and other innovations from this new crowd would be redundant. We did not need another bureaucracy in Washington to make our task more difficult. However, during the next two decades all of us interested in foreign aid slowly recognized that this early environmental arrogance of the AID masked fundamental flaws in promoting near-term economic betterment with little regard to the long-term sustainability of development efforts.

At the time, the AID leadership readily acknowledged that international development programs had serious problems. Several dams financed by the AID in Africa had changed river patterns, creating huge pools of stagnant water that had become expanded breeding grounds for waterborne diseases. But the Agency rationalized this unfortunate side effect as a small price to pay for electricity and irrigation. The AID was assisting Brazil penetrate the Amazon jungle. Plant varieties were being destroyed, and new diseases were being introduced to isolated populations which had not developed natural immunities to many of the diseases that plagued modern societies. But the alternative was for the population of northeast Brazil to continue to live in abject poverty, an alternative that the Agency rejected. The AID had discovered pollution havens in the developing world where dirty industrial plants of multinational companies, alongside even dirtier plants of local industry, did not have to abide with regulations limiting the discharge of contaminants nor with guidelines for disposal of wastes which they would face in the United States. Nonetheless, these countries had few alternatives for satisfying market demands and for creating jobs in the near term, and the foreign companies were contributing relatively little additional pollution to the already high contamination levels, concluded the AID experts.

In short, we at the AID were overly confident that our specialists who had toiled in developing countries for more than two decades adequately appreciated the importance of environmental protection and clearly knew far more about the side effects of international development than did any of this new breed of environmentalists. The Agency could easily refine its documents and expand its programs for training specialists from the poor countries, so we thought, to show compliance with both the letter and the spirit of the new law which clearly seemed targeted at domestic agencies. In fact, we wondered how these other agencies had not thought seriously in the past about the environmental impacts of their programs in the United States.

I did not have to argue our case that the new environmental offices of the government should forget about the AID. The young political enthusiasts at the end of the table were kept busy confronting a large audience of cynical bureaucrats from the old-line departments of government. The newcomers had their hands full with an agenda designed in the first instance to reign in the river development projects of the Bureau of Reclamation, the coastal engineering activities of the Corps of Engineers, and controversial pork barrel construction programs around the United States. Also, the senior diplomat from the Department of State made an impassioned plea that international activities were too complicated to be encumbered by Environmental Impact Statements. He urged that foreign policy efforts concentrate on encouraging the nations of western Europe to be as forward-looking as the United States. All agreed. Thus, for the next few years the new environmental movement in the United States had little influence on foreign assistance programs.

During these early days of the modern environmental era, my colleagues and I at the AID heard constant rumblings from the White House Office of Science and Technology (now named the Office of Science and Technology Policy) over trace levels of toxic chemicals being found in the environment. However, these poisonous materials were considered to be by-products of American industry, and industrial facilities were few and far between in the countries where the AID had programs. Meanwhile, agricultural specialists of the AID had become very defensive over criticism of their programs to ship large quantities of pesticides to developing countries. They received a number of complaints from officials in the developing countries concerning the emphasis the AID was placing on the use of pesticides. In response, the AID specialists blamed the poor training of pesticide applicators in those specific countries while dismissing the possibility that the real problem was inherent in the heavy reliance on pesticides to raise agricultural production levels.

Within a few years, private environmental groups in the United States and abroad began to understand the weaknesses in the orientation of almost all foreign assistance programs toward near-term economic payoffs from specific projects, with longer-term environmental impacts only a secondary concern. In the United States, they laid at the AID's doorstep every environmental insult they could uncover in the Third World. They had found many, including the buildup of pesticides in the deltas of Central America, loss of soil productivity through excessive use of fertilizers in Pakistan, increasing air pollution in cities throughout Asia, and uncontrolled acid runoff from mining operations into the streams of Africa. As a result, during the 1970s the U.S. Congress began wrapping its appropriations for the AID in environmental blankets. Before using appropriated funds, the AID must now ensure that environmental side effects would not outweigh the benefits of the programs. With its budget at stake, the AID quickly responded with many environmental pronouncements and programs.

Today, the AID professes to be an environmental agency. Forestry preservation, land reclamation, reduced use of pesticides, and pollution control technologies have become highly visible components of the AID lexicon. However, as discussed in a later chapter, neither the AID nor the policy agencies of the U.S. government have adequately recognized the central issue that is relentlessly degrading environmental quality in developing countries and is also becoming a dominant concern on the global environmental scene. This issue is the spiraling growth of the world's population.

The Environmental Movement Restores the American Spirit

In addition to the enactment of the National Environmental Policy Act, three other events highlighted the entry of the United States into the environmental age at the beginning of the 1970s. In April 1970, the first Earth Day was celebrated in Washington and in many other communities throughout the United States by "the little old ladies in tennis shoes" and by hundreds of thousands of others in "the biggest street festival since the Japanese surrender in 1945." The EPA was established later that year. Finally, in 1972 the United Nations organized a major environmental conference in Stockholm. No longer would there be unconstrained economic development at home or abroad.

Many historians attribute the environmental awakening in the United States to a series of alarming environmental events that were easily understood by the man on the street and by the politicians in Washington. They point to Rachel Carson's book *Silent Spring,* which was first published in 1962, describing the impact of pesticides on wildlife and on humans.[1] For more than a decade, and indeed even to today, this book has attracted a wide readership which has uniformly reacted with alarm to the revelations over the side effects of pesticides. Also, historians usually recount the serious air pollution problems that resulted in a shutting down of industry in Birmingham, Alabama, and temporarily paralyzed the city, while across the country Los Angeles was continuously reeling from smog alerts. Meanwhile, stress the history books, state officials indicted mercury as destroying the fishing industry in the western part of Lake Erie. In Cleveland, the pollution entering the Lake became so thick that the Cuyahoga River caught fire. Finally, they note, the oil slick from a breakaway oil well off the coast of Santa Barbara blemished one of the nation's most beautiful coastal areas and disrupted migratory routes for whales and other marine mammals.

With regard to establishment of the EPA, a former White House aide has described the internal debates within the Nixon Administration as follows:

> At cabinet meetings, HEW Secretary Bob Finch, responsible for air pollution controls, and Transportation Secretary John Volpe argued over which department should take the lead in developing a research program for unconventional low-polluting automobiles. On pesticides, Walter Hickel at Interior and Finch argued for tighter pesticide controls while Agriculture Secretary Clifford Harden emphasized the increased crop productivity resulting from application of pesticides. And Secretary of State Bill Rogers weighed in expressing concern on whether a ban on DDT in this country might restrict the supply of DDT to the developing countries. Hickel, who at the time handled

> water pollution control over at Interior, wanted more money for sewage treatment control; Bob Mayo, director of the Bureau of the Budget, would have none of it. Maurice Stans at Commerce was wary of tighter pollution controls and what effect this might have on corporate profits. Paul McCracken, Chairman of the President's Council of Economic Advisers, worried that we would be uncompetitive in international markets if our product prices reflected the costs of pollution abatement standards that were more stringent than those of other countries. There was hardly a cabinet officer who did not have a stake in the environment issue. Even the Postmaster General joined the debate, offering to use postal cars to test an experimental fleet of low-pollution cars.
>
> The cabinet meeting left President Nixon dissatisfied. There was no overall strategy, too many unanswered questions. Should enforcement be done by regulation, or by user fees, or by a combination of both? What were the overall costs to industry and the consumer in terms of the increased price of products resulting from various pollution abatement schedules under varying standards and regulations? Finally, what would the various cleanup scenarios do to the federal budget? Nixon clearly needed a pollution czar and one agency to look for the answers.[2]

The EPA was born shortly thereafter.

The environmental shocks of the 1960s were certainly traumatic, and the bureaucratic confusion in Washington was deplorable. However, there were other important dimensions to the emerging enthusiasm and commitment for a higher quality of life and new environmental priorities at both the national and local levels. First, the American economy was on a roll despite the drain of the Vietnam war. Fueled by spectacular technological achievements throughout the industrial complex, business was prospering at home and abroad. The United States could afford the costs of environmental regulations. Probably of even greater significance, the country was ready for programs that were new and exciting, but most importantly for a cause which was respectable and good.

Several personal experiences brought home to me the depths to which our country had fallen in the 1960s and early 1970s and the importance of the environmental movement in helping to pull society out of a swamp of despair. In 1965 I landed at the Los Angeles airport near the flames of the Watts race riots, and at major intersections national guardsmen sitting in jeeps—behind machine guns mounted on tripods—directed me to my home in the center of the city. Three years later in 1968 after leaving my office in the center of Washington with Fourteenth Street ablaze from another wave of race riots, I drove by army troops crouched behind the same type of machine guns on the White House lawn. Then, in 1971 I traveled to Saigon and to the highlands of Vietnam to develop programs of technical aid and cooperation only to find myself constantly surrounded by bodyguards in jeeps with these all-too-familiar machine guns.

Throughout this period I watched from vantage points both inside and outside the White House as our national leaders agonized over how to calm racial tensions and how to extract the nation from the mire of Vietnam. All the time, they searched for ways to unify the country in other areas where a national consensus could help offset the divisiveness of racism and war. By the end of the 1960s the United States, as a nation and as a people, was ready for new alternatives, alternatives rooted in goodness and designed to benefit all people. Environmental protection was one alternative.

From Eyesores and Odors to Toxics and Cancer

The increasing revelations of public health hazards resulting from the improper handling of toxic chemicals, and particularly pesticides, clearly contributed to the new emphasis in Washington on environmental protection during the late 1960s and early 1970s. However, with the EPA established and many agencies hard at work preparing Environmental Impact Statements, the attention of the government and the public quickly turned to other more visible environmental problems. Everyone wanted prompt action to clean up the algae clogging lakes and ponds, to reduce smog and ozone hanging over cities, to move municipal sewage causing environmental eyesores and foul odors, to restrict uninhibited land development laying waste to bountiful marshlands, and to redirect garbage to more suitable dump sites.[3]

Toxic chemicals were not totally neglected. They simply were not at the top of the priority lists of regulators in Washington or in the state capitals. Hundreds of chemicals being used by industry were of potential concern, warned scientists; but the harmful chemical effects, when they were present, were hidden from view. Also, the agricultural lobby, working through the Congress and the Department of Agriculture, ensured that pesticidal chemicals would remain one of the backbones of American agriculture and would not become a victim of overzealous regulators within the EPA.

The laws for water pollution and air pollution were among the earliest legislative authorities to call specifically for the regulation of toxic chemicals. However, the EPA was uncertain how to assess the hazards from nearly imperceptible levels of mercury, cadmium, lead, and other heavy metals, let alone even more minute amounts of the complicated organic chemicals which were appearing in food and drinking water with increasing regularity. Developing approaches that would limit the discharge of traces of these chemicals into the environment to "acceptable" levels without being excessively demanding of industry was a new type of challenge which the EPA was not eager to tackle.

The Agency engineers argued, with justification, that as a first step the EPA was already requiring the removal of particulates, oil, grease, and other common pollutants from waste streams of industrial facilities throughout the country, and the same pollution control devices that were used to remove these common pollutants would also filter out heavy metals and organic matter. However, while helpful, these devices were not designed to remove all trace amounts of metals and organic chemicals which were of concern in scientific circles.

Meanwhile, several environmental officials housed in a new advisory body in the outer offices of the White House complex—the Council on Environmental Quality which had been established by congressional action—concentrated their efforts on developing an approach which would require the EPA to regulate many "toxic substances." They used this term to encompass all man-made chemicals regardless of the dangers they posed except pesticides, drugs, cosmetics, food additives, and radioactive materials since these categories of chemicals were already regulated under special laws which had been in place for many years. Included among the many unregulated chemicals which attracted their interest were highly reactive chemicals used as intermediate materials in industrial processes: petroleum products; solvents used in dry cleaning establishments as well as more powerful solvents used in metal working facilities, in paints and dyes, and in plastics and artificial rubbers; and a variety of household products such as lyes, varnishes, and cleaning powders.

In 1971 the White House aides began drafting new legislation. They pointed out that the water and air pollution control laws could not adequately regulate all toxic substances that were causing problems: in some cases the need was not simply stopping the discharges of waste chemicals from industrial plants but rather banning chemical products in order to protect workers or consumers. Also, information on the possible health and ecological impacts of many chemical substances was very limited. Therefore, the advocates of new legislation proposed that industrial companies be required to conduct laboratory tests on many chemicals which they sold or planned to sell in order to demonstrate whether or not they posed environmental hazards.[4]

The environmental problems attributed to polychlorinated biphenyls (PCBs) that burst onto the scene in the early 1970s greatly strengthened the case for toxic substances legislation which would "fill the gaps between existing laws." This chemical—which is really a mixture of chemicals with closely related molecular structures—was widely used in the electrical industry. The fire resistant properties of PCBs made them ideal as a coolant for transformers. Transformers are located not only outdoors where we commonly see them next to electrical generation and transmission facilities, but they are also hidden from view in subways, electrical trains, office buildings, and other indoor facilities. In addition, PCBs were important materials for capacitors and other electrical devices used in every city and town.

However, once PCBs escape into the environment, they are nothing but trouble. Unfortunately, they were escaping on a nationwide scale. Leakages and spills had become common during the filling and repair of transformers and capacitors. Also, such equipment often ended up in landfills and junkyards where PCBs drained into the surrounding land and water.

PCBs are highly persistent in the environment: they simply do not break down into other chemicals and go away. They spread through the water, through the air, and through the soil as they withstand the chemical batterings of nature. Originally they were known to poison wildlife and to cause chloracne on people's skin. Then they were shown to cause liver function abnormalities. Now they have been linked by some scientists to cancer and to reproductive disorders although other scientists dispute these contentions.

In the early 1970s PCBs were showing up everywhere—in fish from the Great Lakes, in streams and estuaries, and in soil samples from around the country. The White House sounded the alarm: the Monsanto Company which manufactured almost all PCBs produced in the United States must "voluntarily" close its plant, and Congress must pass new legislation to safeguard the country from other dangerous chemicals in the future.

Monsanto responded and shut down its PCB production facility in East St. Louis. At the same time, PCBs became the "toxic classic" which was repeatedly cited during the next few years in the clamor for new regulatory authority to control toxic chemicals. "What if Monsanto had stonewalled the appeals of the government and not stopped production?" hypothesized the environmentalists. Clearly, tough new regulations were needed, they concluded.

Despite the scare over PCBs, Congress was not very enthusiastic in its initial response to still another major piece of environmental legislation. No one on Capitol Hill understood the magnitude of the proposal to subject all chemicals to regulatory scrutiny. The chemical industry warned that the costs of such legislation would be enormous, potentially affecting annual sales in the tens of billions of dollars. Thus, congressional staffers, together with a few officials of the executive branch, began an educational process that lasted for five years until the new legislation became a reality.

Searching for Legislative Rationality: The Toxic Substances Control Act

When I arrived at the EPA in early 1973, I learned that my principal task was to lead the effort of the executive branch of the government for enactment of the Toxic Substances Control Act. The Nixon Administration had already introduced proposed legislation into the Senate and the House of Representatives. But progress was stalled on Capitol Hill at that time as the education of the interested congressional staffs had progressed very slowly.

During the next three years this legislative lull turned into a series of firestorms as toxic problems erupted in every important congressional district and as concerns over cancer resulting from exposure to chemicals in the environment became inextricably linked with the call for a new toxic substances law. Confusion, confrontation, and in the end cooperation reverberated through the halls of Congress, the offices of the EPA, and the confines of the Office of Management and Budget. Many government departments, state agencies, industrial firms, environmental groups, labor unions, and scientific societies discovered a new legislative initiative that would permanently transform their ways of doing business. I quickly joined the chorus of the original architects of toxic substances legislation in proclaiming that the new legislation was to be different from the earlier environmental laws. It was to be "rational," we boasted.

The basic concern was clear. Life depends on aggregations of chemicals in different forms, and as our standard of living rises, so does our dependence on chemicals. While a few man-made chemicals may at times cause problems, mankind has survived living with synthetic chemicals from the earliest days of alchemy. Since no population is prepared to turn back the clock and return to natural settings, control of chemicals on a massive scale surely must be approached with some care.

Thus, an overriding provision in the toxic substances legislation was a requirement that any regulatory action under the law must be justified by the EPA as providing environmental benefits that outweigh the economic costs of the action.[5] This approach differed from the provisions in the earlier air and water pollution laws for regulating toxic chemicals. According to these earlier laws, determinations of the health and environmental hazards of chemical discharges were the sole consideration as to the appropriateness of regulation. Economic impact had not been a relevant concern.[6]

A second important principle was that the tens of thousands of chemicals already on the market were presumed to be safe unless proven harmful by the EPA. However, many EPA officials were not comfortable with the Agency accepting this burden of proof. They worried that the EPA would not be able to segregate the hazardous chemicals from those that were environmentally benign and then make sufficiently persuasive arguments about the chemical hazards. Also, they feared that imprecise scientific evidence used to justify regulatory action would not stand up when challenged in the courts. Others argued that if the government with all of its resources could not make strong cases, the hazards could not be very serious to begin with.

In contrast, under the law regulating pesticides at the time, the EPA had begun to assert on the basis of limited information that a variety of pesticides already in use throughout the country were hazards. The Agency planned to suspend continued use of these pesticides unless the manufacturers could demonstrate that they were safe. This challenge by the EPA of the safety of pesticides which had been in use for many years was called "rebuttable presumption" of hazard: industry had the responsibility to rebut the presumption of the EPA that the targeted pesticides posed unacceptable risks to human health or to ecological resources.

It was successfully contended by myself and others that such an approach, which shifted the burden of proof from the government to the manufacturer, who was then required to demonstrate that a chemical already in commerce was safe, was not to be followed under the new toxic substances law. The government, and not industry, should have the burden to prove the case of acceptability or unacceptability for chemicals which were already in use. The overwhelming majority of chemicals in use were not causing problems, and therefore chemicals that were already on the market should be considered innocent until shown to be harmful, we repeatedly argued.

However, the most contentious aspects of the proposed new law were the provisions for controlling the introduction into the marketplace of the 1000 to 2000 newly developed chemicals each year. To the environmental lobby, at least initially, the obvious approach was a requirement for approval by the EPA of each new chemical before it could be manufactured. Such a requirement would parallel the EPA's approval process for each new pesticide.

But most parties, including the important environmental groups themselves, soon rejected this approach for several reasons. The pesticides program of the EPA had developed a poor reputation for lengthy delays in reviewing and approving applications for new pesticides, and that program considered only a few dozen new chemicals annually. Imposing comparable delays on more than 1000 new chemicals each year under toxic substances legislation would wreak havoc on the marketing activities of many chemical companies. Uncertainties in marketing provoked by such delays would in turn place a serious chill on research and development activities at a time when the international competitive edge of the U.S. chemical industry rested on a high level of innovation in going from the laboratory to the marketplace.

Also, pesticides are designed to be toxic so they can kill bugs and weeds, and they are deliberately released into the environment. In contrast, toxicity is not a design criterion for industrial and consumer chemicals. Very few of these chemicals are deliberately spread into the environment. The exceptions include highway deicing chemicals, fertilizers, and colorants for golf courses. In short, I, and many others, vigorously argued that the regulatory approaches to new pesticides which are toxic by design and the approaches to other new chemicals which were not designed to be toxic should be different.

Thus, the concept of "premanufacturing notification" evolved. This approach called for companies to report to the EPA all new chemicals which they planned to produce prior to beginning production. The EPA would have a limit of 90 days to uncover health or environmental hazards. The Agency would have the opportunity to review new chemicals, but industry would not be penalized by delays in the review process.

Another fundamental principle embraced by Congress related to the responsibility for testing chemicals suspected of posing environmental hazards. While the burden of proof to demonstrate conclusively that a chemical was an environmental hazard and therefore should be restricted fell to the EPA, Congress decreed that the manufacturers of suspected chemicals should provide the EPA with sufficient information about the chemicals to permit the government to adequately assess the associated risks. Thus, the burden of testing chemicals fell to industry.

Manufacturers had long provided information to the EPA under the air and water pollution laws about many of the chemicals which they produced and about their by-products, but the information requirements of the new law were far more explicit and wide-ranging. In particular, the EPA could require manufacturers to carry out toxicology studies costing more than one million dollars per chemical to develop data concerning the safety of chemicals specified by the Agency. Also, the EPA could order the manufacturers to submit detailed information of the quantities of the chemicals being produced, the by-products of individual production processes, and the present and potential uses of the chemicals.

The new law was to address other knotty issues. Congress decided to exempt small businesses from some regulatory requirements. But we who were to be responsible for implementing the law wondered, "Are chemicals manufactured by small businesses less dangerous than chemicals produced by big companies?" The EPA would protect trade secrets while also honoring requests from the public for information about the chemicals under the Freedom of Information Act. "But won't some groups want the secret data?" we asked. Overlaps with other laws were to be sorted out on a case-by-case basis with an emphasis on using the most cost-effective laws to address specific problems. "But how can any regulatory program admit that its approach is not cost-effective?" Finally, the new law was to encourage efforts of the states to develop complementary regulatory programs. "But does this mean that there will be layers of laws?" Thus, the proposed law seemed to raise more issues than it resolved.

A continuing series of highly publicized toxic problems around the country from 1973 to 1976 focused congressional attention on the proposed toxic substances legislation. Asbestos, vinyl chloride used in plastics, the fumigant ethylene dibromide, and other chemicals were repeatedly showing up in the environment as health or ecological hazards. A common reaction to such incidents was a call by politicians and by the press for new legislation, specifically the Toxic Substances Control Act, even though laws that were already in place may have been quite adequate to address some of the problems.

The incident which imparted the final impetus for the passage of toxic substances legislation involved the chemical Kepone. In 1975 a very small company operating in a converted gasoline station in southern Virginia was producing this highly toxic chemical. The company flushed Kepone residues from the process directly into the James River with devastating impacts on the local fishing industry since the fish became contaminated at levels that were considered unsafe for human consumption.

As a pesticide ingredient, Kepone was explicitly excluded from regulation under the proposed new law. Few congressional leaders understood the differences in the reaches of the pesticides and water pollution laws, which were the basis for cleaning up the Kepone mess, and of the pending toxic substances legislation. They simply wanted some type of action to prevent another Kepone incident. Passage of the Toxic Substances Control Act was a convenient outlet for their frustrations over the government being saddled with a large bill from dredging Kepone off the bottom of an important tributary of the Chesapeake Bay.

Meanwhile, during the mid-1970s cancerphobia reigned in Washington. Every week the press featured scientific findings which suggested that still another chemical was a carcinogen—a chemical that could cause cancer. The government prepared a collection of maps of every state showing the incidence of cancer in each city and county, described as a cancer atlas. The maps suggested correlations between the occurrence of cancer and concentrations of industry, again prompting the Congress to call for new legislation.

A few scientists who had become strong advocates of greater governmental regulation of chemicals made irresponsible statements that ninety percent of cancer deaths could be attributed to environmental chemicals. These statements were echoed by the press. What they apparently meant to say was that 90% of cancers could be traced in part to "environmental factors" including smoking, infections, diets, alcohol consumption, and other sources that had little to do with man-made chemicals escaping into the environment.[7] As the obsession with chemicals and cancer intensified, we in the EPA also lost some of our perspective. We decided to "go with the flow" and ride the wave of cancer warnings. We picked up the chant and began to press for toxic substances legislation by urging that "chemicals be tested for cancer in the laboratory and not tested on people."

Already by the end of 1974, enactment of the new legislation had seemed a certainty. The chemical industry had accepted the inevitability of such legislation. Many industrial leaders believed that prompt enactment of the proposed legislation was better than the uncertainty of future legislation. Most environmental leaders were excited by the prospect of a new law which would provide another tool for forcing industry to become more responsible.

But near the end of 1974 opposition arose from a most unlikely source, the Sierra Club. While the environmental groups had been among the strongest supporters of the legislation from the outset, Sierra Club lobbyists decided that more Democrats were likely to be elected to Congress for 1975–1976 than were in office during 1973–74, and therefore legislation which placed more severe strictures on industry could be expected if enactment were delayed. With the weight of such an influential group against immediate passage of the legislation, congressional support sputtered, and the law was not passed in 1974. The version that emerged two years later was almost identical to the draft opposed by the Sierra Club, which had made a serious miscalculation about the impact of political partisanship on this legislation.

As the debates continued in the congressional committees during 1975 and 1976, concerns over the economic impact of the law arose time and again. Industry, the EPA, and the General Accounting Office, which is an arm of the Congress, carried out hasty economic analyses. The estimated annual impact of the law on industry ranged from $200 million to $2 billion.[8] However, the difficulty in predicting industrial responses to such far-reaching legislation and the uncertainty in estimating lost marketing opportunities became quickly apparent. In particular, the costs in lost sales of products which were not developed due to the chilling effect that the law might have on research activities were impossible to estimate accurately.

In any event, while the economic dimensions of the proposed law were important, they were clearly overshadowed by congressional concerns over PCBs, Kepone, and other problem chemicals, concerns which carried the day and led to enactment of the new law in October 1976.[9]

Then and now, Congress and the American public have perceived toxic chemicals as a threat to public health which must be harnessed, whatever the cost. However, as we will see, once the cost implications in terms of jobs, modified life-styles, and increased taxes from rigid control of large numbers of chemicals become clear, the attitudes of many self-proclaimed environmental zealots rapidly adjust to the realities of the modern industrial age.

A Passive President Accepts the New Toxics Legislation

When the Toxic Substances Control Act finally overcame all the hurdles in the Congress and arrived at the White House for signature, President Ford delayed his response to the very limit of the time allowed until a pocket veto would take effect. Only a few hours remained when he finally signed the legislation into law in the privacy of his office in October 1976.

But why did he delay? Why didn't he celebrate the passage of this complicated legislation with a signing ceremony in the rose garden, a beautiful environmental setting just outside his office? Environmental and industrial leaders, congressional and cabinet officials, and labor and foreign trade executives who had repeatedly clashed during the incubation of the legislation could have joined hands in an unusual display of national consensus. President Ford's campaign for reelection desperately needed demonstrations of support from all sides. Yet the White House limited its public relations effort to a routine press release of the type used to record less-than-noteworthy presidential actions.

Perhaps the president's staff correctly judged that the new legislation was politically insignificant in comparison with education, welfare reform, crime, and other social and economic issues confronting the nation. Perhaps the impassioned congressional debates, the intensity of the lobbying efforts, and the media accounts of the cancer crisis had exaggerated the threat of toxic chemicals. But isn't legislation that could affect the daily workings of tens of thousands of industrial firms throughout the country of considerable importance?

The White House inner circles wanted the chemical problems which regularly punctuated the *Washington Post* and the *New York Times* to go away. After struggling until the last minute to understand the complicated provisions which Congress had crafted, they undoubtedly concluded that the new legislation would help, even though it could hardly be considered monumental in their eyes.

Meanwhile, the Manufacturing Chemists Association (now renamed the Chemical Manufacturers Association) and the Chamber of Commerce had advised the White House of their desires to have the legislation signed and thereby end the regulatory uncertainty that had hung over the chemical industry for five years. The Sierra Club, the Natural Resources Defense Council, and other environmental groups assured presidential aides that their concerns had been addressed. The AFL-CIO argued that safety was even more important than jobs and that the law should be enacted. The Office of Management and Budget, the EPA, and other government agencies endorsed the legislation. Even some of the states had become vocal advocates for the law which would shift to the federal government many of the politically charged problems of public health threats that they were encountering every day. Since no one seemed to object, any need for the president to be concerned with the fine print must have appeared very unimportant in comparison with the larger issues of the reelection campaign. In any event, the president clearly missed a useful opportunity to enhance his image as a responsible advocate of environmental protection.

The Education of the Revolving-Door Regulators

In 1976 only a handful of people, perhaps ten, concentrated in Washington had a good understanding of the scope of the new law and its implications for environmental protection, for commerce, and for the nation's administrative and legal systems. And their understanding was elementary at best. During the years that followed, a very large cadre of specialists in toxic substances control has developed. However, the political leadership of the program has changed frequently, bringing many newcomers into the picture.

Despite sizable research and regulatory programs, the collective understanding within and outside the government of how this legislation can best protect society from toxic insults has advanced very slowly. During the 1970s the key policy issues related to the methodologies needed for assessing risks to health and ecological resources posed by seemingly small amounts of man-made chemicals and for determining when risks were significant. Then, how could such risks be balanced against the costs of regulatory actions, and how could the likely reductions of the risks from such actions be estimated? These same issues constitute the bulk of today's regulatory and judicial agendas, and they are at the heart of the continuing educational process for regulators in Washington and in the state capitals.

The new Toxic Substances Control Act was to be the centerpiece of the government's efforts to regulate toxic chemicals. It would require industry to provide information on the production of chemicals and on the toxicity and behavior of the chemicals in the environment. This information was needed by the EPA to determine whether to impose more stringent controls under a variety of laws on toxic pollutants being discharged into the air and into waterways. And it would allow the EPA to halt or limit manufacturing and marketing activities when necessary.

Thus, the new statute would place a cap on industrial activities that contributed to chemical hazards in the environment. It would reduce the frequency of incidents of harmful chemicals entering our food and water supplies. The law would lead to safer working conditions within factories and safer living conditions in the neighborhoods around large manufacturing complexes.

Finally, the law would provide a new type of administrative framework for reaching environmental decisions. Such decisions would take into account not only environmental hazards but a broad range of economic and even social considerations as well.

When the president signed the law, enthusiasm was high within the EPA, other interested federal agencies, industrial organizations, and environmental groups. The specialists who had devoted five years to working out the intricacies of the law were prepared to translate concepts into action—to end an era of regulatory uncertainty and to begin "balanced" regulation. At the top of the list of immediate actions were three urgent issues.

First, great strides had been made during 1974, 1975, and 1976 to encourage the phasing out of PCBs and to begin cleaning up areas where PCBs had been stored or discarded. Some of the actions had been required by state laws or were carried out in response to federal water pollution regulations. Industry had undertaken other actions on a voluntary basis. Now it was necessary to codify restrictions on PCBs in all-encompassing federal regulations to ensure the permanent shackling of this ubiquitous nemesis.

Second, American scientists had argued persuasively that discharges into the atmosphere of Freon and similar refrigerants called chlorofluorocarbons were degrading the belt of ozone which serves as a radiation shield for the Earth. The new law broadened the legal basis for requiring a reduction in the manufacture and use of these chemicals, and appropriate regulations to limit the production of these chemicals had been prepared. Thus, the first steps could be taken to address the problem of depletion of stratospheric ozone.

Finally, the law called for the EPA to review the properties of all newly developed chemicals before industry began selling the chemicals and to place limitations on the production and use of any new chemicals which posed environmental or health threats. In order to determine whether a chemical was "new," the Agency needed to prepare and publish a master list of all "old" or "existing" chemicals—the chemicals already in production. Any chemical not on the list would then be considered a new chemical. EPA experts had prepared draft regulations to promptly collect from industry information concerning the existing chemicals so the list could be prepared and the review of new chemicals could begin.

However, plans for early action in these and other areas soon fell apart. The Carter Administration arrived in Washington. Not unexpectedly, that Administration brought its own team to the EPA to administer this law as well as the other environmental laws. Unfortunately, some appointees mistakenly thought that only Democrats were the real advocates of the environmental movement and that the Republicans who preceded them had been dragging their feet on environmental issues. Based on my experience, however, environmental partisanship traceable to political allegiances had few policy footholds within the EPA under the Republican administrations of the 1970s. In any event, the Democratic appointees slowed down activities which had been started by their Republican predecessors at the EPA, intent on placing their own fingerprints on future regulations. Many regulations which were in various stages of development were placed in limbo.

As to my personal involvement, the new Administrator of the EPA removed me from my career civil service position as the Director of the EPA's Office of Toxic Substances. Apologizing all the while, his assistant told me the reason behind my transfer to another position: I knew too much about the intricacies of the new law, and it would be difficult for less knowledgeable political appointees to take control of the program if I remained in my position. I now appreciate the depth of mistrust that frequently divides political appointees from career technocrats in Washington. Particularly in the environmental field with all the complexities of scientific uncertainty, incumbent career officials frequently are perceived as having too many opportunities to drive policy in directions which are consistent with personal convictions but which may not correspond to the views of political leaders.

The new team started over. Many had impressive credentials, and they recruited a large number of well-qualified specialists from within and outside the EPA to manage the toxic substances program. However, these new leaders did not appreciate the complexities of addressing tens of thousands of chemicals of different characteristics, with different chemicals touching different interest groups.

Fortunately, the three sets of regulations mentioned above—phasing out PCBs, limitations on chlorofluorocarbons, and preparation of the master list of existing chemicals—had reached advanced stages, particularly in the eyes of the Congress and public interest groups. Thus, the new team had to truncate its self-education to a few months, and original regulatory timetables slipped by only six months to two years in these three areas as the new Administration tried to understand the new law. In many other areas, however, the Agency's implementation of the law during the years of the Carter presidency can only be described as minimal, with original timetables slipping many years.

As one example of misdirected efforts, on several occasions the EPA spent hundreds of thousands of dollars over many months and even years developing and defending regulations which would require industry to conduct laboratory tests on single chemicals that were suspected of having toxic properties. The Agency became so deeply enmeshed in the details of the regulatory process that it forgot the purpose of the law—namely, to have many chemicals tested as quickly as possible. The Agency could have simply carried out the tests itself at lower costs and saved two to three years in the process. Preferably the Agency could have developed regulations to require testing by industry of groups of chemicals rather than repeating the regulatory process for each chemical. Agency specialists had advocated that approach since 1973. The value of grouping chemicals for regulatory purposes was finally recognized toward the end of the 1970s, and the EPA eventually adopted this approach.

By the time the Carter Administration had gained a firm grasp of the new legislation, the Reagan Administration arrived in Washington. Again, a new set of administrators entered the EPA and other agencies with related interests. Suspicious of regulation in general, and particularly regulation aimed at manufacturing activities, this new team applied the brakes to regulatory actions which they judged might have adverse economic consequences. They argued that the scientific basis for many of the regulations which were in the process of development within the EPA was faulty, but at the same time they reduced the research budget of the EPA by 30%—the very budget intended to improve the scientific basis of regulations. By the time the Administration had begun to appreciate the benefits which might be derived from the toxic substances law, the Congress had enacted a new mechanism for addressing many of the central concerns over toxic chemicals—the Superfund legislation to clean up abandoned chemical wastes. Much of the attention concerning the assessment of problems associated with toxic chemicals rapidly shifted to programs required for implementing this new legislation.

The most successful aspect of the Toxic Substances Control Act has been the EPA's process of reviewing premanufacturing notifications by industry which signal commercial intentions to market new chemicals. After a slow start, the EPA has done a fine job in establishing procedures which quickly separate notifications for chemicals that are environmentally benign from notifications that raise genuine health and environmental concerns. A surprisingly few of the 2000 notifications received each year fall into the latter category, reflecting a heightened awareness within the industry that new chemicals simply must be environmentally acceptable or the EPA will restrict them.

With regard to curbing existing chemicals which present environmental problems, however, the EPA's record under the Toxic Substances Control Act has been quite controversial. In 14 years the Agency has taken actions to limit only a handful of chemicals and to require testing of only a small percentage of chemicals identified by experts as potentially toxic.[10] Environmental groups contend that many chemicals identified in the early 1970s as being environmental hazards should have been limited long ago—some dyes, plastics, solvents, and heavy metals, for example. Others argue that on close examination the alleged hazards of the 1970s were overstated and that there has been no need to take more aggressive action to restrict large numbers of chemicals which have been safely handled for decades. While the debate will continue as to whether some of the common chemicals which have been used for many years pose a health or environmental threat, the EPA sorely needs to give higher priority to its efforts to screen the potential hazards associated with many existing chemicals. The Agency has been preoccupied with refining and re-refining regulations to control PCBs, supervising the removal of asbestos from schoolrooms and public buildings, and restricting chlorofluorocarbons. The Congress has ordered all of these actions, and of course they are important.

But the EPA needs a broader surveillance effort to prevent harm from the relatively few hazardous substances among many chemicals that are bought and sold every day.

Impatience Leads to Alternative Routes for Controlling Toxics

While most political leaders had focused on the proposed Toxic Substances Control Act during the mid-1970s as the eventual solution to the toxic "threat," environmental activists in Washington had become restless with the slow pace of the legislative process. Many scientific societies and private environmental organizations were not content in limiting their interests in toxic chemicals to lobbying for legislation. At the same time, the philanthropic foundations which had provided financial support for private environmental groups to serve as counterweights to industrial lobbies were considering whether to reduce their funding in view of the expanding role of the EPA. The environmental organizations needed dramatic evidence of their continuing value in keeping the EPA on course and thereby serving as an environmental conscience of the nation. Thus, several environmental groups were looking for ways to force the EPA to greatly accelerate its timetable for controlling toxic chemicals under legal statutes which were already in place.

In one of the most significant activities of the environmental groups during the 1970s, the Natural Resources Defense Council claimed that industry was irresponsibly discharging toxic chemicals from plants in every state and county with devastating effects on rivers, streams, and estuaries. The council identified over 125 chemical pollutants of primary concern and turned to the courts to require the EPA to take regulatory action to limit these discharges. The judge was sympathetic. Within a few months the EPA and the environmentalists signed a Consent Decree, binding the EPA to begin controlling the 125 chemical pollutants. During the decade that followed, these chemicals became well known as the "priority pollutants."[11]

Limiting discharges of chemicals into aquatic resources as required in the Consent Decree was not completely new to federal or state governments. Since its inception, the EPA had issued guidance to the states on water quality "criteria." This guidance set forth chemical contamination levels—or criteria—which were environmentally acceptable in streams and rivers that were sources of drinking water or irrigation water or were holiday retreats for boating, swimming, and other recreational purposes. Many state agencies used the EPA guidance in determining appropriate restrictions on pollutants in an effort to keep water pollution below the recommended levels. Initially, the EPA developed criteria levels for metals (e.g., mercury, cadmium, lead, tin) and a few pesticides (e.g., DDT, dieldrin, aldrin). As the pressures from environmental groups increased, the Agency established criteria levels for other chemicals as well.

Also as a part of the national effort to protect the quality of water resources, in the early 1970s the EPA had begun a long and tedious process of requiring industry to install improved technologies for removing pollutants from wastewater streams in all types of industrial facilities. The costs to industry were substantial, and in many cases industry challenged the reasonableness of the new requirements in court actions which resulted in frequent delays of many months or even years in the promulgation of final regulations. During the years that followed, the EPA identified hundreds of appropriate technologies for limiting pollutant discharges, and industry installed many of these technologies in thousands of facilities. These technologies removed both biological and chemical pollutants from waste streams. But as previously noted, they were not designed to eliminate many of the minute traces of toxic chemicals which were so much more potent than small amounts of other pollutants.

The Consent Decree for the priority pollutants required substantial research efforts that stretched the limits of our understanding of how to deal with very small quantities of toxic chemicals. The EPA established new programs to refine the estimates of the health effects and ecological hazards of each of the 125 chemicals on the list. The Agency developed, validated, and then proscribed analytical chemistry methods for measuring low concentrations of the chemicals. Researchers tested new engineering technologies to determine their effectiveness in removing trace chemicals from different types of waste streams. The priority pollutants surely awakened the environmental community to the practicality and to the costs of controlling toxic discharges into the environment.

Providing impetus for the Consent Decree were many newspaper accounts of chemical contaminants entering drinking water supplies—in Cincinnati, in Philadelphia, in Southern California, and elsewhere. As analytical techniques for measuring very small levels of chemicals improved, chemists began discovering traces of all types of pollutants in drinking water. Scientists attributed some of these contaminants to natural runoff from areas where trees and other plants had decayed. Other chemical pollutants were by-products of the process of chlorinating water to destroy biological contaminants. Still other unwanted chemicals seemed unequivocally linked to discharges from industrial facilities and from sewage treatment plants.

In one highly publicized account, scientists from the University of New Orleans identified many organic compounds in the tap water in Louisiana—compounds which had been linked with cancer in experiments with laboratory animals. They then attempted to show that the gastrointestinal cancer rates in Louisiana were abnormally high, thus implying a cause-and-effect relationship. The results of this investigation and similar studies in another half-dozen cities at about the same time fell far short of demonstrating that the drinking water was not safe or that there was a statistical correlation between cancer rates and specific contaminants in the drinking water. But the scientists captured the attention of both congressional and administration leaders who soon became committed to strengthening still another law, the Safe Drinking Water Act aimed at reducing trace levels of toxic impurities as well as other types of contaminants in water supplies.[12]

Meanwhile, EPA scientists began conducting regular nationwide surveys of pollutants in drinking water. The results have sometimes been alarming as more and more trace chemicals appear throughout the nation. Debates punctuate many scientific meetings over whether trace amounts of different chemicals, individually or in combination, pose a health risk. In any event, as Congress tightened the law governing drinking water contamination, the list of chemicals subject to regulation increased substantially from the nine metals and pesticides that had been controlled since the early 1970s. By 1990 the number of chemicals for which acceptable levels had been adopted was about three dozen with limits for 50 more to be established by 1992.

Toxic air pollutants also became increasingly important in the 1970s and 1980s. Of course, emissions from leaded gasoline had been a national concern for many years as evidence continued to mount about the effect on the learning capabilities of children from lead that is inhaled or injested. Also, as discussed in some detail in the next chapter, the early regulations under the Clean Air Act 20 years ago placed explicit limitations on air emissions from industrial facilities of mercury, asbestos, and beryllium, and requirements for removal of particulates from air emissions also reduced discharges of many chemicals. Benzene emissions soon became a particularly contentious issue since this chemical permeates every gasoline station. Vinyl chloride was added to the list of controlled air emissions, and in the 1980s attention turned to discharges of radioactive contaminants. Soon thereafter a list of 37 additional chemical pollutants, primarily carcinogens, became the center of heated debates concerning the approach for controlling "air toxics."[13]

During the 1980s, repeated surveys by the EPA throughout the country showed that air toxics are present at some level in almost every industrial area. However, only now has a major effort to reduce air toxics moved to the top of the priority list of federal and state regulators. The proposed 1990 revision of the Clean Air Act will, if enacted, at last require stringent measures to reduce toxic emissions. No longer will leaky valves and pumps be tolerated. Timetables for retrofitting old plants will be accelerated, and limitations on emissions from new plants significantly reduced. Several states are still not satisfied with action in Washington and have enacted more stringent legislation. Such legislation advances the regulatory timetable in those states by several years and requires a reduction of toxic air emissions of up to 50% by the early 1990s.

But, it was the sudden emergence of the problems of hazardous waste disposal which truly brought the issue of toxic chemicals onto center stage. Abandoned wastes were uncovered seemingly everywhere beginning in the late 1970s and early 1980s: for example, in the Valley of the Drums in West Virginia, in abandoned pesticide mixing areas in California, and even at shopping malls in New Jersey. And no one wanted to accept responsibility for cleaning them up. The wastes contained thousands of different types of chemicals which were leaking into the groundwater, seeping into the atmosphere, and even exploding as they were disturbed. Government inspectors and the press found toxic chemicals in rusty barrels, in abandoned ponds, and in mounds of dirt. Just determining which chemicals were present became a Herculean task.

In response the Congress designed the Superfund legislation to provide a comprehensive framework for setting priorities and cleaning up hazardous wastes. The intent was to force responsible parties to pay the bill whenever possible. A trust fund—a Superfund—was established to be used for cleanups when the original dumpers could not be clearly identified or when emergency situations required immediate action. The trust fund receives money from taxes on crude oil and major commercial chemicals. Initially the state governments were supposed to pay 10% of the costs to the government at privately owned sites and 50% at those that were publicly owned. However, the ground rules for state contributions have been under frequent review and revision. Another goal of the Superfund program is to advance technological capabilities in all aspects of hazardous waste management, treatment, and disposal.

When the legislation was originally enacted, no one fully appreciated the enormity of the problems. More than 30,000 potential Superfund sites have been identified over the years. Although the EPA has already categorized a significant number as too minor to warrant Superfund status, the number of remaining sites of genuine concern is still overwhelmingly large.

The EPA also began discovering large quantities of hazardous chemicals in municipal landfills across the country along with garbage and other household wastes. Then in the mid-1980s we learned that hundreds of thousands of buried storage tanks containing heating oil, gasoline, solvents, and other types of chemicals were corroding. Some were already leaking and many more would eventually be threatening groundwater resources. The Congress enacted additional legislation to address these problems within the framework of the Resources Conservation and Recovery Act. The contamination of the land had reached unprecedented levels and continues to this day.

The costs of cleaning up abandoned wastes are enormous. The costs of safely disposing of the refuse are staggering. The costs of constructing and managing proper waste sites to avoid similar problems in the future are also very high. The costs of containing municipal landfills which punctuate every town and village are yet to be estimated.

The assessment, containment, and cleanup of hazardous wastes tax the limits of understanding of every aspect of the science and economics of toxic substances. The problems faced by the EPA and the nation in implementing the Toxic Substances Control Act and the toxics provisions of the Clean Water Act, the Clean Air Act, and the Safe Drinking Water Act are formidable. But they seem small compared to the problems of cleaning up the toxic chemical debris simply dumped upon society in now abandoned waste sites by past generations.

Ecological Concerns Return to the Top of the Regulatory Agenda

The previous discussions have emphasized the dangers, both real and imagined, to human health from toxic substances. During most of the 1970s the public and the government agencies were preoccupied with the threat to people from inadequate control of toxic chemicals. But such a public health orientation, often to the neglect of broader environmental concerns, was not anticipated by many of the leaders of the environmental movement in the early 1970s who were concerned with disruption of ecosystems and destruction of flora and fauna by uncontrolled pollutants.

Indeed, ecology had been on center stage in Washington and around the country when the nation entered the current environmental era in 1970. Ecologists were concerned, and indeed outraged, over the impact of chemical pollution on natural resources. They witnessed fish kills near the discharge pipes of industrial plants in Tennessee. They deplored the contamination of shellfish in the Chesapeake Bay. They saw the forests disappear near the lead smelters of Missouri. They grimaced as wildlife breeding areas became chemical waste beds along the coasts. They reinforced the earlier protest of Rachel Carson in condemning the excessive use of pesticides which were disrupting nature's protective mechanisms in agricultural areas.

However, as new environmental laws were passed and cost–benefit arguments dominated discussions of strategies for carrying out these laws, most of the attention of the regulators and the press turned away from ecology to issues of public health. No one questioned the importance of protecting nature's resources. Nonetheless, protecting lives seemed a far more urgent matter. Human suffering has always been front-page material, and impending death and disease certainly have great impact in arguments over larger federal budgets. By 1975 protecting human health was the dominant environmental issue with preservation of the ecology considered a secondary concern. In many ways, the EPA had become a health agency.

No words provoked greater reactions on Capitol Hill than "cancer," "birth defects," and "genetic damage." It seemed that almost every chemical that was leaking into the environment suddenly had the potential for causing the most dreaded form of human agony, according to some environmental extremists. Laboratory scientists were discovering tumors and deformed offsprings in their experiments with rats and mice with increasing frequency. Reports reaching the government and the public of industrial workers suffering from cancer were on the rise. Government agencies published long lists of chemicals which showed carcinogenic tendencies in tests conducted in scientific laboratories in the United States or abroad. Court litigation targeted at recovering costs for pain and suffering due to exposure to environmental chemicals was on the increase. Medical doctors had replaced ecologists as the most important experts on the risks associated with chemical pollution.

By 1980, however, ecological problems had returned to the mainstream of environmental debates. Of course, attention to the health effects of chemicals, and particularly cancer, continued. But ecological hazards such as acid rain began to emerge as major problems. Lakes in New England that had been biologically alive were becoming sterile, and healthy forests bathed in ozone and other pollutants in several parts of the country were losing their foliage. The sport fishing industry was suffering, and timber sales were threatened.

Still, the worries were far more profound than the declining contribution to the economy of fish and timber from a few lakes and forests. Americans who lived in New England or who regularly visited the impacted areas persuasively argued that the eventual effects of permanent acidification of the countryside could not be predicted and that a price tag could not be attached to the long-term losses of ecological resources. Furthermore, America was experiencing a politically charged phenomenon: the activities of people living in one part of the country were destroying ecological resources in other regions hundreds or even thousands of miles away. In particular, the power plants of the Ohio Valley, using low-cost coal from the Northwest, were indicted as the principal cause of the death of New England lakes. Who should pay the bill to arrest the problem?

A few years later, even more ominous ecological concerns arose. Scientists found a hole in the stratospheric ozone belt over Antarctica. This hole was directly linked to the continued use of chlorofluorocarbons. Also, scientists theorized that we were becoming encased in a greenhouse constructed by chemical pollutants, and the hot summers of the late 1980s seemed to provide the evidence they needed to document their theories. Today scientists clearly recognize that global ecosystems are changing due to human activities. Because of heating induced by air pollutants, low-lying coastal areas may be flooded during the next century, agricultural lands parched, and skin cancer rates increased, some warn.

Society needs as never before the expertise and dedication of highly skilled ecologists who appreciate the complex webs which weave together man's synthetic habits and the surrounding ecosystems. Ecologists led the surge of interest 20 years ago in promoting greater awareness of the interactions between man and nature. They raised critical questions, and now they must be key players in the search for answers that currently elude us.

The Many Dimensions of the Toxics Problem

As we have seen, during the past two decades synthetic chemicals have been discovered in trace amounts in every conceivable part of our environment. Even the air in some American living rooms contains formaldehyde, and the milk of many nursing mothers reveals minute levels of pesticide residues. The sediments in some rivers have become repositories for toxic metals, and pine needles are being discolored under the weight of chemical fallout from cars and factories. These fingerprints of human activity are the inevitable consequences of our industrialized culture.

Neither ecologists nor other types of scientists know the extent of the risks posed to humans or to natural resources by the very low concentrations of chemicals which we encounter every day. Most Americans assume that these risks are trivial in comparison with the risks from smoking and from poor nutritional practices. Some of us may gain distorted comfort in knowing that timber losses from fire and parasites are a more immediate problem than losses from acid rain and that projects to divert rivers for agricultural purposes may be greater threats to the water supplies of many communities than trace levels of chemical contaminants. But the bottom line is that no one really knows the seriousness of the problem of toxic chemicals in the environment, and chemical contamination will clearly be one of the central political and scientific issues of the 1990s.

As discussed, the regulatory responses to concerns over chemicals in the environment have been manyfold, and in many instances pollution emissions have fallen. More than one dozen federal laws with explicit provisions for controlling chemical pollutants are on the books. Every state has general environmental legislation which encompasses chemicals, and several have special legislation to limit toxic chemical releases into the environment. Hundreds of general regulations are now in effect, and hundreds of thousands of regulatory permits have been issued to individual facilities for controlling discharges of chemicals. The country has taken great strides in developing the procedural framework for controlling chemicals during the 1990s.

Many institutions have responded to chemical problems in a number of ways. Mortgage companies which finance homes near waste sites, insurance companies with industrial customers, and wholesalers who handle agricultural products from contaminated areas have adjusted their policies and their prices to cover their increased risks. Industrial companies and municipalities have invested enormous financial resources in retrofitting polluting facilities and in installing new facilities which pollute less. Waste minimization has become the byword of the environmental movement.

Still, by any measure, progress in containing potentially harmful chemical contaminants has been modest. As each trash truck goes to the landfill and as each smokestack releases emissions, the environmental burden of man-made chemicals increases. Many chemicals quickly decompose into harmless products, but others do not. In many localized areas, and even on a global basis, the capabilities of streams or the atmosphere to dilute pollution to acceptable limits and to continuously absorb synthetic chemicals have been exceeded.

In 1989 the nation received a jolt when the public learned that industry was still discharging billions of pounds of chemicals into the environment every year. These releases into the air, water, and land are all within legal limits. "What was the purpose of all those laws, regulations, and permits anyway?" many Americans ask.

Leading up to the revelations in 1989, the EPA had responded to a new law by undertaking a systematic survey of industrial discharges of chemicals. The results were surprising to many cities and towns when the principal offenders were identified. While confirmation of Texas and Louisiana as the locations of the heaviest discharges was to be expected, few would have guessed that Salt Lake County in Utah was among the ten counties most affected. Environmentalists were quick to point out that within the 7.5 billion pounds of legally emitted chemicals were large quantities of 39 known or "probable" carcinogens. Industry questioned the relevance of aggregated data, arguing that since the discharges were so widely dispersed, the pollution levels at any single location were well below the thresholds of concern.[14]

The press coverage was extensive, and it certainly promoted the belief that toxic discharges posed a significant threat. For example, under the headline "The Many Uses of Toxics—and Dangers They Pose To You," *USA Today* described the health effects of 60 of the most prevalent chemicals included in these discharges, including the following five "common" chemicals with discharges exceeding 100 million pounds per year:

> Toluene: Solvent used in preparation of perfumes, medicines, dyes, explosives, detergents, aviation gasoline, and other chemicals. Hazard: highly flammable and explosive; toxic by ingestion, inhalation, and skin contact.
>
> Sulfuric Acid: In fertilizers, chemicals, dyes, rayon, film; widely used by the metals industry. Hazard: strong irritant.
>
> 1,1,1 Trichloroethane: Solvent for cleaning precision instruments; also in pesticides, textiles. Hazard: irritating to eyes and skin.
>
> Methyl ethyl ketone: Solvent in making plastics, textiles, paint and paint removers, adhesives. Hazard: flammable, explosive; toxic by inhalation.
>
> Dichloromethane: Industrial solvent and paint stripper; in aerosol and pesticide products; used in photographic film production. Hazard: carcinogen.[15]

As indicated, the press identified the hazards associated with each chemical but failed to note that the chemicals must survive in the environment and be present at sufficiently high levels to do damage.

These chemicals are not confined to a few locations but are present in many areas of the country. Clearly, a better characterization of their pervasiveness, together with assessments of their concentrations in the environment, is now in order to effect rational control of the sources of undesirable pollutants.

Controlling Chemical Pollution in the 1990s

What are the challenges of the next decade when the American public will have no alternative but to live with an even heavier burden of environmental chemicals? The chapters which follow begin with a critique of the past approach of trying to control chemicals one by one and with suggestions for future approaches to cope with thousands of chemicals.

This book emphasizes the changing trends and the outlook for the 1990s. We will explore the central issue of risk with all of its uncertainties. The popular misconception of the ease in separating the science of assessing risk from the art of controlling risks is highlighted. Also, better communication between government agencies and the public about risks and corrective actions is underscored as deserving much higher priority in the future. The public doesn't want simply to listen. It is demanding to be heard.

A chapter on the chemical waste problems confronting the nation suggests modified approaches to handling the growing pyramids of debris accumulating around us. Individual states are now becoming more important focal points for controlling chemicals at waste sites, in rivers and streams, and in the atmosphere. Some of their past and current efforts deserve greater attention.

This book investigates the schism that has developed between the regulators and the regulated industry. This divisiveness must be tempered if real progress is to be made in cleaning up the environment.

Environmental problems are no longer simply national problems. They cross international frontiers with increasing regularity. Thus, the global setting for the national and international environmental debates during the 1990s is given high priority in the discussions.

In conclusion, we ask whether science and technology will come to the rescue and cut the costs of cleaning up the nation.

This book attempts to help set the stage for more aggressive action by government, by industry, and by citizens to limit damage to our nation and the world from the use of chemicals that have raised the standard of living everywhere. If each of us is willing to make modest economic sacrifices, we can enter the 21st century on a wave of hope and optimism that the quality of life as we have known it will continue.

2 Controlling 3000 Chemicals
One by One

One hallmark of contemporary America is the short life span of its crises. A problem emerges suddenly, builds swiftly to crisis proportions, briefly dominates public consciousness and concern, and then abruptly fades from view.

—Wall Street Journal

The number of potentially toxic substances is enormous. The number of commercial chemicals is 70,000. . . . Distressingly little is known about the toxic effects of many chemicals.

—The Conservation Foundation

One part per million: One drop of a toxic chemical in a barrel full of drops of pure water. One part per billion: One fine grain of a toxic chemical in a large bathtub full of grains of clean sand.

—An environmental scientist

The Toxic Chemical of the Month: Hexachlorobenzene

In 1973 I had been on the job at the EPA for less than one week when telephone calls from Louisiana started jamming the switchboard. The Governor and his aides challenged our competence. The state legislators gave us a piece of their minds. The farmers demanded redress.

Inspectors of the Food and Drug Administration (FDA) at a slaughterhouse in Texas had refused to certify shipments of beef originating in an area near Geismar, Louisiana, as "safe." In turn, all slaughterhouses in the region began turning back the Louisiana cattle. But the farmers had nowhere else to sell their cattle.

During routine chemical analyses of beef samples from the Louisiana herds, government scientists had detected the chemical contaminant hexachlorobenzene. They measured contamination levels of 0.1 to 6 parts per million (ppm). Since 0.3 ppm would have been reason enough to reject food according to the FDA, beef from Geismar was labeled as unsafe. But the FDA had established its "action level" of safety of 0.3 ppm several years earlier before the EPA had entered the scene. Now the responsibility for determining a safe level fell to the EPA, not the FDA. On the one hand, a bombardment of telephone calls to the EPA demanded that the Agency take steps to clean up the source of contamination. On the other, farmers challenged the original FDA action level as excessively conservative. They urged a new and more lenient level of safety.

Hexachlorobenzene was being discarded as a powdery waste from chemical plants. In this case, a large chemical complex of the Dow Chemical Company near Geismar had produced the noxious by-product. Cattle normally grazed in the fields alongside the many chemical plants punctuating the landscape. Trucks from a local company hauled the waste from the plant, along with wastes from many other plants in the area, to a landfill about 25 miles from the Dow plant. However, the haulers sometimes neglected to cover their waste loads during the trips so the powder blew across the fields en route to the disposal site. The operator of the landfill was also lax and simply piled the fluffy waste on top of other wastes without bothering to spread dirt on top of each load. On windy days the powder filled the air and drifted onto the fields. Twenty thousand grazing cattle had become contaminated from the powder.

One of the immediate tasks at the EPA was to determine a safe level for traces of hexachlorobenzene in food. Government inspectors urgently needed an acceptable level which they could use for certifying the safety of beef. At the same time, the Agency had to take action to put an end to the pollution of the countryside.

Agricultural officials in Washington, D.C. and Baton Rouge urged as high an inspection level as possible to minimize the rate of rejecting cattle, since each head was worth about $500. As might be expected, the FDA was skeptical that any change in its old level of 0.3 ppm was warranted. The EPA Administrator decided to seize upon this incident, and specifically the establishment of a safety level, as a highly visible demonstration of the Agency's sensitivity to the need to balance health concerns with economic considerations.

Meanwhile, the Dow Chemical Company pointed its finger at the waste hauler and the waste site operator. In Washington, Louisiana politicians were demanding special congressional legislation that would compensate the farmers for their losses.

The effects of hexachlorobenzene on humans were not well known at the time and are uncertain even now. In the 1950s a number of people in Turkey had accidentally ingested large quantities of the chemical with serious illnesses and a few deaths resulting. Several limited studies subsequently carried out in research laboratories with experimental rats suggested that human ingestion of the chemical over a long period of time at levels above 10 ppm could lead to serious illnesses and perhaps death in a manner akin to poisoning. However, no information was available concerning the possibility of deleterious effects of low levels in the range encountered in Louisiana.

In setting the original action level of 0.3 ppm, officials of the FDA had simply adopted a level used for other chemicals with similar molecular structures which scientists had studied more intensively. Ironically, the same FDA officials argued with great passion during discussions with the EPA that similar structures did not necessarily mean similar toxicities. Meanwhile, the World Health Organization had developed a guideline of 1 ppm as a safe level for the chemical. However, in probing the history of this international decision, specialists from the EPA discovered that the guideline was derived from the judgment of a group of medical experts who had met in Geneva. They had relied primarily on intuition rather than on documented scientific studies since there were no solid studies.

EPA staff members tried to trace the distribution of beef from slaughterhouses in Texas to consumers throughout the country. The idea was to show that by the time the Louisiana beef appeared on the dining room table, it would be so intermingled with other beef from uncontaminated areas that no individual consumer would repeatedly receive portions of meat with high traces of hexachlorobenzene. Therefore, the safe level might be higher than otherwise indicated. This analysis proved fruitless. While supermarket chains are excellent dilution mechanisms through the normal mixing of small cuts of meat from various sources, there was always the possibility that families who ate large amounts of meat would buy and freeze entire sides of beef from Louisiana.

Predicting how different safety levels would affect financial losses was even more speculative than estimating the health consequences of allowing different levels of contamination to pass safety inspections. Louisiana agricultural specialists had collected and analyzed hundreds of samples of flesh from the cattle in the field. Therefore, the EPA had a good profile of the levels of contamination within the herds and could estimate the number of cattle with contamination above 0.3, 0.5, 1, 2 ppm, and higher levels which were being considered for the action level. However, some of the cattle had been switched to clean feed, and EPA scientists did not know how to estimate the rate at which hexachlorobenzene would be flushed out of their internal systems. Thus, they had no basis for predicting how many cattle would have to be destroyed at different levels of safety. Also, neither Dow nor the state agencies would indicate the extent of economic relief, if any, which they would provide the farmers until the EPA established the action level. Finally, farmers always display remarkable ingenuity in cutting their financial losses when faced with economic disasters.

After several weeks of heated interagency debates in Washington, vociferous public meetings in Louisiana, and raucous sessions between EPA officials and the Louisiana congressional delegation, the EPA Administrator decided that 0.5 ppm was the appropriate level. Two considerations were pivotal in selecting this number. First, it was substantially below the guideline recommended by the experts assembled by the World Health Organization a few years earlier and therefore would appear to reflect the caution of the EPA when lives were at stake. Second, it was the highest number which the FDA would support regardless of financial losses. The views of the FDA were very important. The EPA was a new government agency trying to establish a sound reputation within a skeptical health community. Being aligned with the FDA, which was known for its caution on health issues, was a highly desirable goal.

Thus, when the time came to make the decision, economic factors were considered, but not very seriously. In retrospect, the downplaying of economic factors was warranted in this instance. Once the action level of 0.5 ppm was established, the dire predictions of economic consequences disappeared. Immediately, farmers put their cattle on clean feed which quickly diluted the hexachlorobenzene concentrations to levels lower than 0.5 ppm in over 95% of the cattle. The state then purchased almost all of the remaining cattle for demonstration farms for retarded children who would observe but not eat the cattle.

This was my introduction to the "toxic of the month" approach that dominated the control of chemicals during the 1970s. As different chemicals were pinned down as the causes of problems in specific parts of the country, the federal government reacted, usually in a responsible manner. Unfortunately, the government seldom anticipated such crises and was almost always guided to problems by public outrage rather than by systematic study of the most threatening chemicals in the environment.

During the past decade, the hunt for toxic chemicals has been more far-reaching. At times it has seemed as if the government agencies have moved from the toxic of the month to the toxic of the day, and the number of chemicals subject to some type of regulatory controls by the EPA has grown to more than 3000. Many are used in pesticides. Many others are embedded in wastes being hauled away for permanent disposal. Still others are emitted into the air and streams at many industrial facilities. And a few are highly valuable chemicals used by businesses or even around the home where they must be handled with care.

During the 1990s the list of controlled chemicals will continue to grow, and the regulations governing how they are handled will become ever more complicated. In some cases, the chemicals will be controlled in groups. Sometimes the regulations will be highly specific to individual chemicals. But can the regulatory agencies, the manufacturing companies, and the interested public responsibly cope with such a large array of pollutants which are often characterized more by their differences than by their similarities? Do the agencies have the intellectual and financial wherewithal to develop hundreds and even thousands of customized approaches to deal with this chemical mosaic?

As we will see, recent history has shown that some chemicals must be singled out for individualized attention. They simply are too toxic, too pervasive in the environment, and too idiosyncratic in their behavior to be easily lumped with other chemicals which are targets for control. History has also demonstrated that alternative approaches are needed to address the vast majority of the chemicals which may cause problems. The past experiences in dealing with individual chemicals which are discussed below should help set the stage for developing new strategies that will reduce risks from many more toxic chemicals in the future.

Containing One of the Most Toxic Pollutants: Dioxin

During the early 1980s, a serious problem erupted with a long-time toxic nemesis—dioxin. Times Beach, a few miles west of St. Louis, suddenly won the title of "pollution capital of the nation" and subsequently turned into one of America's first toxic ghost towns.

The story began in 1969 when a small firm named North Eastern Pharmaceutical and Chemical Company took over a manufacturing plant in Verona, Missouri. Earlier the company had produced Agent Orange, a mixture of the plant-killing herbicides 2,4,5-T and 2,4-D, which was widely used as a defoliant to clear the vegetation on the battlefields in Vietnam. This chemical mixture was frequently contaminated with small amounts of dioxin. At the same time, Agent Orange was used in abundance, and the total amount of dioxin contamination was substantial.

Dioxin contains a wide variety of highly toxic molecules which have very similar structures. While toxicity varies from molecule to molecule, all are hazardous to some degree; and they are usually considered as a single pollutant, at least in the eyes of the public. Some of the very potent effects of dioxin, and particularly its genetic impacts, have been of concern for many years although the scientific debates continue to today over the extent that American army troops were exposed in Vietnam to dangerous levels of Agent Orange laced with dioxin. Indeed, several detailed studies were unable to attribute any health effects among American soldiers to the use of Agent Orange.

In 1969, the North Eastern Pharmaceutical and Chemical Company began producing another toxic chemical, hexachlorophene, for use in pesticides and cosmetics. This production process also generated substantial quantities of dioxin as a by-product. Until 1971 the company shipped its waste which was laden with dioxin to an appropriate chemical disposal facility. But then in order to save money the company contracted with a small disposal firm—Independent Petroleum. This firm in turn subcontracted with a waste hauler from Missouri, Russell Bliss. His job was to take away oily sludge impregnated with dioxin which had accumulated on the bottom of the chemical mixing containers.

Bliss was not content simply to receive money for disposal of the waste—he adopted a new version of recycling. He sprayed the oily residues along the dirt roads and byways of Missouri in return for fees from local officials and residents who had discovered a low-cost, dust suppression program. In the early 1970s Bliss, in addition to spraying roads, coated several dusty horse arenas with his newly acquired oils, and hundreds of animals soon became sick and died. The state authorities sent soil samples from the horse arenas to scientists at a federal laboratory in Atlanta who found dioxin in the oil at levels far greater than the concentrations reported from the use of Agent Orange in Vietnam. Meanwhile, the attention of the federal and state agencies was diverted by the discovery of still higher levels of dioxin within the North Eastern manufacturing site at Verona. Prompted by such scrutiny, the owner of the plant began to treat the waste at the site at his own expense and successfully reduced the hazard. In the absence of additional reports of immediate problems, the issue of dioxin waned in the mid-1970s.

Several years later in 1979, dioxin again made headlines as the EPA began receiving new reports of dioxin contamination throughout Missouri. The agency launched a major investigation and began reconstructing the history of the dioxin problem in detail. Through an extensive sampling program, the EPA discovered particularly widespread dioxin contamination in Times Beach, a small community along the Missouri River.

Greatly exaggerated rumors of the effects of Agent Orange quickly spread throughout the community. Environmental sampling teams in protective gear resembling moon suits moved in next to children playing in backyards. The residents demanded that the "government" buy their homes so that they could move to a safe place. During the state election campaign in the fall of 1982, dioxin contamination took its place as a major political issue. A few politicians touted new chemical and biological sprays that would neutralize dioxin—an unrealistic claim. Most simply demanded compensation for the affected neighborhoods.

At that time, the extent of the health effects from exposures to low levels of dioxin could only be inferred from previous investigations of the effects of dioxin in more massive doses. For example, following several industrial accidents a few years earlier which released dioxin at high levels into the atmosphere, some of the exposed workers had developed chloracne on the skin, a general sense of fatigue, disturbances in nervous system responses, and enlargements of the liver. Though these conditions seemed to recede after a few years, scientific studies with laboratory animals indicated the possibility that dioxin also caused birth defects and perhaps cancer at levels which were not much above the highest contamination levels in Times Beach. Meanwhile, medical studies of the residents of Times Beach uncovered no significant health impacts of dioxin.[1]

Again, as with the case of hexachlorobenzene and the contaminated cattle, the key issue was to determine the level at which contamination was unacceptable. What level of dioxin in soil makes an area uninhabitable? Using as the point of departure results from studies of birth defects caused by dioxin ingested by monkeys, government toxicologists in Atlanta concluded that a contamination level of 1 part per billion (ppb) was the dividing line between habitable and uninhabitable. In their calculations they estimated the extent that dioxin would cling to soil, the likelihood that children playing outside would lick their dirty hands, and the possibility that the soil would be suspended in the air and breathed by residents.

Once the media had trumpeted the conclusions of the health researchers, the debate over habitability ended. For Times Beach residents, the implications of this health finding were clear. Since levels considerably higher than 1 ppb had been discovered in several areas of town, they were convinced that evacuation was the only course of action.

Who could challenge the views of medical experts employed by the federal government? They clearly preempted the responsibility of public officials to consider the social and economic implications in reaching an environmental decision, since residents simply would not live in areas pronounced by researchers to be unsafe. Buried in the pronouncement of the scientists was their decision that the concentration level at which dioxin might affect people should be 1000 times lower than the level that affected laboratory animals in order to provide an appropriate "margin of safety" for society regardless of the cost implications.

Finally, in February 1983 the EPA announced that it would buy the contaminated property where levels exceeded 1 ppb at prices reflecting land values before the contamination. After delays of some months, the federal government purchased the property, the residents moved out, and Times Beach was permanently cordoned off as a monument to the laxity of a chemical producer and the greed of a waste hauler who himself was soon carted off to jail.

The discovery of dioxin in Missouri triggered one of the biggest nationwide hunts for a single pollutant in history. Around the country, a major mission of each of the EPA's ten regional offices was to search out every conceivable source of dioxin. For several years the hunt continued. The EPA directed special attention to facilities which at one time had produced Agent Orange. Other pesticide production and processing plants were also implicated. In addition, the Agency targeted waste disposal areas for attention.

Soon dioxin was discovered to be a by-product of many waste incinerators. Indeed, some everyday combustion processes release trace amounts of dioxin into the air. And municipal incinerators which burn large quantities of plastics or wood products preserved with chlorophenol can produce significant dioxin emission. The implications of these findings were ominous, and the level of restrictions to be imposed on incineration of wastes remains a highly controversial issue.

Returning to soil and water contamination not only at Times Beach but also in many other parts of the country, the level of 1 ppb proscribed by the toxicologists becomes very important. The cleanup costs might be substantial even though the contamination is only slightly above the proscribed level. If the contamination level is slightly below 1 ppb, lack of action could mean the continuation of a health risk even if cleanup costs were very low. The researchers should not have been so quick to pronounce such a definitive level of safety based on limited scientific data—a level which can have tremendous financial and social implications for all concerned. As discussed in the next chapter, both researchers and policy officials must become more skilled in dealing with the uncertainties of risk evaluations and less prone to make definitive assertions as to the dividing line between safe and unsafe.

In summary, dioxin is one of the best examples of how the government has focused enormous resources on a specific environmental chemical. Its toxicity has caused anxieties wherever it has been uncovered, and its historic association with Agent Orange has heightened political interest in its immediate containment. Now, as additional laboratory studies are reported, scientists are beginning to question whether the hazard of dioxin has been exaggerated.[2] However, to back off on the stringency of limitations which have been codified is politically very difficult for the government regardless of new scientific findings.

Setting National Standards for Toxic Air Pollutants

In many communities during the past two decades, the public has become aroused by dioxin, PCBs, hexachlorobenzene, and other toxic contaminants which threaten their well-being. However, the federal agencies began many years earlier to pay attention to a handful of other pollutants which could pose hazards for workers and for neighborhoods abutting industrial facilities. By the early 1970s, the EPA in particular had adopted a chemical-by-chemical approach to setting pollutant discharge standards as a cornerstone of the government's efforts to protect public health from toxic exposures.

Since that time, the EPA in very workmanlike fashion has identified hundreds of problem chemicals. The Agency has studied the likely health and ecological effects of each of these chemicals should they escape into the environment. It has estimated the extent people and natural resources are being exposed to individual chemicals, or will be exposed, if the substances are not contained. The Agency also considers the feasibility of various ways of controlling each chemical and then decides whether to attempt regulation of a particular chemical on a national basis. A good example of this chemical-by-chemical control method is the early regulatory action taken within the framework of the Clean Air Act.

In April 1973 the EPA limited emissions into the atmosphere of asbestos, beryllium, and mercury from industrial and other facilities. The Congress explicitly required that health concerns should take precedence over cost implications when considering hazardous air pollutants, and these three chemicals were the first to be addressed. Though scientific data as to their health effects were far from definitive, the EPA nevertheless imposed tight restrictions on emissions into the atmosphere.

First, with regard to asbestos, scientists had established a clear association between exposures of shipyard workers to asbestos and a higher-than-expected incidence of bronchial cancer. Also, they had shown that asbestos caused cancer of the membrane and lining of the chest and abdomen. Nonetheless, they had not been able to relate the amount of asbestos entering the body to the likelihood of cancer. Moreover, techniques for measuring levels of asbestos emissions from plant stacks had not been developed. Despite these uncertainties, the EPA developed an environmental standard that limited "visible" emissions from asbestos facilities. In short, an asbestos cloud—however faint—in the sky indicated a violation. Also, the EPA specified the types of technology which must be used to remove asbestos from emissions in order to reduce even the invisible emissions.

Turning to beryllium which is used in nuclear facilities as well as in other branches of industry, the U.S. Atomic Energy Commission had already established in 1949 0.01 microgram of beryllium per cubic meter of gases as their limit on emissions from plants under contract with the government. This limit was based on concerns over earlier reports that beryllium caused fibrosis and possibly cancer of the lungs of industrial workers. During the following 25 years, there were no reported cases of diseases related to beryllium among the populations living near these plants. The National Academy of Sciences reviewed the available information on the toxicity of beryllium and concluded that the original standard represented a safe level of exposure. Therefore, the EPA adopted the early limitation as the standard for all industrial facilities and, in effect, simply codified existing practices throughout the country in handling a potentially hazardous chemical.

Finally, with regard to mercury, the amount that is breathed must be considered together with trace levels that may contaminate food or water. An expert group convened by several U.S. government agencies in the early 1970s analyzed a number of episodes of mercury poisoning in Japan and concluded that 4 micrograms of methyl mercury per kilogram of bodyweight per day would result in the poisoning of a sensitive adult. They suggested that an exposure of about 30 micrograms per day for an "average" person weighing 70 kilograms or 154 pounds would be acceptable. This level would provide for a tenfold safety factor to compensate for the uncertainties in the scientific data being used. In addition to considering poisoning, the experts kept in mind the possibility of damage of the fetus.

At that time, a usual diet, together with drinking water, probably resulted in the injestion of 10 micrograms of mercury per day, according to the experts. Thus, in order to restrict total intake to 30 micrograms, the average intake from air would have to be limited to 20 micrograms. If we assume the quantity of air inhaled through normal breathing as 20 cubic meters per day, the air could contain an average daily concentration of no more than 1 microgram of mercury per cubic meter. This level served as the basis for the standard.[3]

Thus, we see how the EPA customized the approach to setting the standard for each chemical depending on the character of the public health threat and the nature of the scientific information that was available.

During the next few years, the EPA slowly expanded its control of a few hazardous air pollutants as more problems were identified. In particular, the Agency established emission standards for benzene, vinyl chloride, arsenic, and radioactive chemicals. Then in the mid-1980s the EPA initiated greatly enlarged nationwide surveys of the extent of the air toxics problem. Specifically, the EPA searched for 45 specific toxic chemicals in urban air. The Agency attempted to estimate the cancer risks, but not other types of acute or delayed effects which are more difficult to identify, associated with exposures throughout the country to these air pollutants. The estimated cancer incidence was about 0.2% of the total national cancer incidence, an estimate that now seems high to many experts. As a result of these investigations, in 1985 the EPA decided that carbon tetrachloride, cadmium, chromium, chloroform, ethylene oxide, ethylene dichloride, butadiene, perchloroethylene, and trichloroethylene were next in line for regulation.[4]

The main problem highlighted by the EPA studies is the difficulty of isolating individual chemicals in the atmosphere as the cause of cancer or any other adverse health effect. There are so many different types of chemicals present in our cities from so many different sources that developing cause-and-effect linkages is usually impossible. Some chemicals seem obviously to be of more significance than others, based on their high levels of toxicity and the relatively large quantities of these chemicals in the air. According to the EPA, risks result in large measure from "complex pollutant mixtures typical of urban ambient air."[4] Thus, the chemical-by-chemical regulatory efforts of the past seem inadequate to address the problems now facing us. At the same time, developing estimates of how risks could be reduced through specific regulatory measures to reduce emissions of individual chemicals or groups of chemicals remains a formidable problem.

As noted in the preceding chapter in 1989 the Bush Administration proposed a new and far-reaching approach to the control of toxic air pollutants. If this proposal is adopted by the Congress, restrictions will be imposed on emissions of over 190 chemicals through an approach that calls for industry to use the most effective technologies for cutting emission levels. The industries responsible for emissions of these chemicals will be required to install technology mandated by the government to limit their discharges, technology referred to as "maximum achievable control technology." For new facilities, such a requirement means that controls must be at least as stringent as the best emissions control already achieved in practice by a similar facility anywhere in the country. For retrofitting existing plants, controls must be as stringent as controls "typical" of the best performing technologies at similar facilities already on-line. Limitations on discharges from those facilities which are in ten categories deemed to be the worst sources of toxics will be in place within two years, and limitations on the other facilities within five to seven years. Once these controls are installed, the government may take further steps to reduce any residual risks from toxic emissions.

This highly desirable change in the Clean Air Act takes the initial steps in avoiding the endless debates of the uncertain health hazards of individual chemicals while relying on the demonstrated capability of industry to reduce hazards without further delay. Controversy will continue over whether there are significant residual risks from escaping pollutants after the new technologies are in place, but much of the emissions problem will have been solved. The experts will never be able to determine with certainty which chemical mixtures are safe and which are hazardous. We do know that chemical air pollutants are seldom if ever beneficial. If some plants are capable of employing stringent controls and still remain in business, then other plants should be expected to perform with comparable efficiency.[5]

These controls are limited to industrial sources which are responsible for only a portion, perhaps 25%, of toxic air pollutants. Sewage plants and electrical utilities, for example, will not be included, and the status of dry cleaning establishments and gasoline stations is unresolved. Separate regulatory action is being directed to motor vehicles which are responsible for up to 50% of the health impacts from air toxics, according to the EPA—benzene, diesel fumes, and other hydrocarbons. A principal initiative in the proposed Clean Air Act in this regard is the call for greater use of clean fuels, and particularly blends of gasoline with methanol and ethanol.

A very serious step has been proposed to control air toxics. Also, the important principle of controlling toxic chemicals in large groups, in this case a group of almost 200 pollutants, using proven technologies has been strongly reinforced.

Regulating Categories of Chemicals

Historically, environmental protection efforts have spawned separate programs to control chemicals which pollute the air, those which degrade water resources, those which show up in solid wastes, and those which are used in drugs, pesticides, or food additives. Thus, chemicals have been placed in groups for regulatory purposes and are subject to different laws according to how they are used or how they reach the environment.

Frequently, however, the same chemicals appear in many forms—in air, in water, in food. Hence, they attract regulatory attention under several laws. In some cases a single chemical may be regulated under as many as ten different laws. The ground rules for regulating a chemical may vary from law to law.

Uncertainties as to how chemicals behave in the environment further complicate the control of chemicals which move with alacrity from environmental compartment to compartment. Will airborne lead contaminate vegetable gardens? Will cadmium in river sediments be stirred up and make fish inedible? Will solvents buried in waste sites vaporize and leak through cracks in the soil into nearby neighborhoods?

Fourteen years ago, the Toxic Substances Control Act was designed as an umbrella law—to bridge regulatory gaps and reduce legal redundancies which too often had prompted governmental stuttering in the control of chemicals that the public may encounter in many different forms. Congressional expectations that the requirements of different laws would be easily meshed through this new regulatory authority have been excessively optimistic. Bureaucracies in general, and regulatory bodies in particular, resist coordination efforts no matter how many details are spelled out in law. Only now, after many years of conflict and confusion as to how chemicals are to be constrained under a variety of laws, have the EPA and the other regulatory agencies begun to integrate their efforts on a broad basis to address many chemicals that fall within different regulatory groupings.

The need for other types of groupings of chemicals became obvious during the debates which led to the enactment of the toxic substances legislation: many pollutants should be regulated in clusters rather than as individual chemicals. More than 70,000 chemicals are used throughout the country. Thousands of chemicals are discharged into the environment as by-products of manufacturing processes. Many more new chemicals are being developed in research laboratories. Hence, total reliance on one-by-one regulation is simply not practical.

Of initial interest were groupings of organic chemicals with similar molecular structures which, it is believed, have similar although not identical toxicity characteristics. Peroxides and azo dyes, for example, were identified long ago as categories of chemical compounds that include many individual chemicals worthy of regulatory scrutiny. While each chemical has unique properties, many chemicals within a single category pose similar problems.

As previously noted, PCBs and dioxin are mixtures of closely related chemicals. While the individual chemicals in each of these mixtures may have somewhat different toxicological characteristics, most of them are harmful if handled improperly. More than one of the constituent chemicals are invariably present, and formidable problems arise in the laboratory in distinguishing precisely among the individual chemicals which make up the mixtures. Thus, trying to develop a different level of control for each of the dozens of constituents is seldom worth the effort.

However, federal agencies have been hesitant to treat chemicals as groups which are based on similarities in their molecular structures. Of course, when nothing is known about a chemical except the character of its molecules, specialists draw inferences as to whether there may be a hazard based on the known toxicity of similar molecules. But these inferences cannot always stand up as regulatory judgments.

As we will see, risk assessment methods have relied on intensive investigations of individual chemicals with discrete molecular structures. Regulatory actions traditionally depend on such assessments, and lawyers are comfortable defending governmental actions which rely on laboratory studies of chemicals one by one. Extrapolating laboratory findings of toxicity from one chemical to another opens up many scientific avenues for challenge.

In several cases, the EPA has addressed groups of chemicals with closely related structures. For example, over the years the EPA has identified general categories of chemicals which include at least some chemicals with toxicities of concern. The EPA then has required industry to carry out specified types of laboratory tests of other chemicals that fall within these main groups in order to determine, one by one, their toxicities. As another example, the EPA reviews the properties of new chemicals which any company plans to produce. As part of the process for determining the need for regulatory action, the government scientists compare the molecular structures of the proposed new chemicals with the structures of other chemicals which have been shown to be either problems or, on the other hand, environmentally benign.

Turning to the effects that toxic pollutants have on humans, no grouping of chemicals has received more notoriety in Washington or throughout the country than carcinogens, or chemicals which cause cancer. For two decades the debates have raged, with cynics arguing that man is not a giant rat who develops tumors as readily as do experimental animals in response to environmental chemicals. Scientists rightfully argue that while laboratory experiments have their limitations, experiments on laboratory animals provide the best approach for clarifying the biological impacts of chemicals other than using human subjects for experimentation purposes.

First a handful, then several dozen, and now several hundred chemicals have been indicted by government agencies as carcinogens. About two dozen of these have been shown directly through studies of their effects on humans to cause cancer in humans with the remainder being of concern due to the potency they have demonstrated in the laboratory. Qualifying words such as "probable" carcinogens or "weak" carcinogens are sometimes included in the indictments to take into account uncertainties. Once a chemical is considered by the EPA or another regulatory agency to show signs of carcinogenicity, various types of regulatory restrictions may take immediate effect. As an extreme, chemicals with even the slightest carcinogenic tendencies are simply banned by law from use in food additives.

Volume of production is another criterion that has been considered for grouping chemicals of high toxicity. Toxic chemicals which are produced in large quantities may be of more concern due to the possibility of accidents during manufacture, shipment, or storage than chemicals produced in very small quantities and not shipped around the country. Also, the Toxic Substances Control Act exempts from some regulatory requirements chemicals produced only in small volumes for research purposes.

Finally, the likelihood of chemicals coming into direct contact with humans or ecological resources provides a strong rationale for grouping these chemicals for regulatory purposes on the basis of "exposure." For example, chemicals which are deliberately dispersed into the environment such as aerosol propellants should be of more concern than chemicals of comparable toxicity which remain contained within sealed vessels in chemical plants. Chemicals which come into close contact with people such as newspaper inks, dyes, and household cleansers have a greater potential for causing problems than those which only reach people accidentally. Also, chemicals used in open areas such as solvents and paints are usually more worrisome than chemicals which are not in continuing contact with the atmosphere.

Regulatory agencies, and particularly the EPA, have considered all of the foregoing approaches for categorizing chemicals. However, the agencies constantly hesitate over the scientific defense of their conclusions if they attempt to address too many chemicals simultaneously. Meanwhile, industry leaders are skeptical of any schemes that would enable the EPA to broaden its regulatory net for capturing more chemicals through grouping or any other approach.

At the root of the restrained approaches of the agencies in trying to limit groups of chemical pollutants is the preoccupation of scientists with individual chemicals during the risk assessment process. They have become used to studying chemicals one at a time. But the public is seldom exposed to chemicals one by one. Approaches to risk assessment which take into account exposure to different levels of many chemicals with different toxicities are much more complicated, but they would be much more realistic. Greater attention to this neglected problem, and specifically greater reliance on epidemiological studies of what is really happening to people in polluted areas, is essential to undergird more effective control of the large numbers of potentially harmful chemicals which are contaminating the environment.

Waste Products Laden with Hazardous Chemicals

The difficulties in addressing chemicals one by one become especially apparent in the control of solid and liquid wastes—wastes often containing dozens of chemical pollutants. As a starting point, the EPA and state environmental agencies have prepared long lists of regulated waste chemicals. These lists encompass over 1000 chemicals including all of the well-known problem chemicals of the past—toxic metals, pesticides, and corrosive acids, for example. Chemicals that leach from plastics discarded by every household are listed, and the waste solvents from dry cleaning establishments are itemized. Many other chemicals found in "ordinary" trash and in the by-products of every municipal sewage treatment plant are included.

The EPA's regulations prohibit the disposal into the ground of several hundred chemicals in order to ensure that groundwater will be protected from the most dangerous pollutants. These pollutants must be incinerated at high temperatures, chemically decomposed, or reclaimed for further beneficial use. Companion regulations specify methods for determining whether other chemicals or specific types of waste streams which contain chemicals should also be considered hazardous. If they are, they are destined for disposal in waste sites meeting specifications for maximum containment. Still other regulations simply require industrial facilities to publicly declare the quantities of certain chemicals that they are discharging into the atmosphere or into streams or are sending to landfills.

The lists of waste chemicals have triggered many actions by the federal agencies, state authorities, industry, and the scientific community. Numerous studies are under way to characterize the toxicity of the individual chemicals and their mobility once they are in the environment. At what concentrations will they harm humans or ecological resources? Will they explode or catch fire at high concentrations? How easily can they move through the environment—through water and through soil? Will they decompose as they come into contact with other chemicals? What materials can be used to contain these chemicals?

Some of these chemicals have been studied by the government for years. Library shelves are piled high with reports of the health and environmental effects of hundreds of toxic chemicals. Indeed, a cottage industry of firms along the beltway of Washington, D.C., thrives on revising studies of individual chemicals for government agencies and industrial organizations which are suddenly confronted with chemicals that have been problems for others in the past.

Most studies are directed toward individual chemicals even though waste products are usually contaminated with many chemicals. However, the problems of hazardous wastes are not simply the sums of the problems posed by individual chemicals. These chemicals can interact with one another. Also, the solutions to hazardous waste disposal problems may not be obvious from studies of individual chemicals, for measures that will successfully contain or destroy some chemicals may be ineffective for controlling others.

Meanwhile, chemists are hard at work developing better, cheaper, and faster methods for analyzing waste products in order to determine which chemicals are present and their concentrations. They have become keenly aware that their techniques must simultaneously identify large numbers of chemicals in the wastes. For example, an analytical technique called x-ray fluorescence permits identification of two dozen toxic metals. Sometimes, however, a highly specialized technique is needed to identify a single chemical such as mercury. Mass spectrometry is a particularly important analytical method. It permits identification of hundreds of organic chemicals including both chemicals which were suspected all along of being present in the waste and chemicals which show up unexpectedly.

If very low concentrations are of concern, such as in the case of dioxin, development of very expensive laboratory techniques to identify and measure the chemical may be required. In 1984, for example, the EPA's Las Vegas laboratory spent $750,000 to purchase a triple sector mass spectrometer for measuring trace quantities of dioxin. Of course the instrument could also be used to find other pollutants as well although at high costs. The real problem, however, was not cost, but rather finding a scientist with enough experience to operate this highly sophisticated instrument which was still a novelty in the analytical chemistry community.

Even when armed with sophisticated assessment tools, scientists face formidable problems in analyzing wastes. Abandoned hazardous waste sites come in all configurations and contain every conceivable mixture of chemicals. Sometimes the history of dumping can be reconstructed, and reasonably good information is available on the contents of the site—at least as it was at the time of disposal. In many cases, however, the contents of corroding drums are not known, particularly if the disposal dates back many years; and the EPA has uncovered many examples of the pumping of liquid wastes of unknown origin into pits which were then covered with dirt. Thus, the laboratory chemist may have little advance warning of what he or she is looking for. Also, before the sample arrives at the laboratory, it may begin to disintegrate. Crude techniques for removing the sample from the waste pile may result in the loss of pollutants or the addition of sample contaminants, and less than perfect techniques for preserving the sample in its original form may result in changes in the sample en route to the laboratory.

A related concern is the huge volume of hazardous wastes in America's 6500 municipal landfills. The volume of trash in these dumps is so large that determining, let alone correcting, the toxics problem at any site seems like an insurmountable task. On a national basis more than half of our municipal trash is classified as paper, paperboard, and yard wastes. Substantial quantities of glass, food, and textile wastes are also present. The EPA has estimated that a community of 100,000 people deposits almost 500 tons of hazardous wastes in addition to non-hazardous waste in landfills each year. These sites were not designed to contain leaks, and the resources of communities to correct past sins are in very short supply.

An immediate problem both at hazardous waste sites and at landfills is to determine whether harmful chemicals are leaking from the disposal areas—leaking into the air, into nearby surface waters, or into the groundwater. Based on knowledge of some of the contents of the site, scientists usually concentrate their efforts on monitoring for a few "indicator" chemicals, chemicals which are known to be present in the waste and which do not decompose as they are washed or blown through the environment. The presence of these chemicals outside the site suggests that other chemicals may be leaking out as well. Highly toxic chemicals are often selected as indicator chemicals since discovery of a potential hazard off-site immediately buttresses the legal case for forcing prompt corrective action and for penalizing the dumpers.

Waste disposal is currently conceived as a problem of chemical containment although we hope that someday destruction of most wastes may become technically and economically feasible. The identification of the hundreds and eventually thousands of chemicals of concern is a sensible first step. If engineers could indeed contain all of the currently identified chemicals for the indefinite future, the hazardous waste problem would be largely solved. However, during the 1990s financial resources are not unlimited, and only modest steps toward controlling large quantities of so many chemicals in so many locations will be possible. These steps should focus on containing the maximum number of chemicals which pose a threat. This strategy both relies on and transcends chemical-by-chemical control.

Let us now turn to a specific example of how my colleagues at the EPA and I attempted to determine the hazards at a waste site by assessing the problems posed by individual chemicals and then by aggregating these findings.

150 Chemicals Told the Story of Love Canal

In June 1980 I was one of four EPA managers whom the EPA Administrator summoned to his office in Washington where we received instructions to carry out a crash study of the safety of a residential area of about 600 homes in Niagara Falls. These homes surrounded a previously evacuated area of 100 homes immediately adjacent to Love Canal. The Congress, the local authorities of Niagara Falls, and many scientists were vigorously criticizing previous studies of the area carried out by the state of New York and several EPA offices: the environmental sampling was faulty, quality control over the laboratory analyses was inadequate, and unwarranted conclusions were drawn from inadequate data, according to the critics. Now the EPA would devote the full resources of its research laboratories to "do it right," so the Agency promised.

The EPA's Las Vegas laboratory, where I was the Director, was responsible for improving monitoring programs at hazardous waste sites throughout the nation. On the airplane flight to Washington, I reviewed our earlier experiences in designing large monitoring programs to assess toxic chemical contamination. I was particularly concerned as to how the field teams would take samples of soil, water, and air around homes occupied by people who were already very upset and scared without further aggravating the situation.

As we assembled on the eleventh floor of the EPA headquarters, the Administrator and the Deputy Administrator were waiting. The Deputy quickly took charge, enunciating a very simple message: bring back the answer prior to the November presidential election as to whether residents of the area should be relocated. The good news was that there would be no political bias in the way our specialists designed and carried out the monitoring study. The bad news for us was that there were no new funds for the program, and we would have to find the money in our laboratory budgets. As we will see, the bad news for the Deputy Administrator, which we broke to her slowly in the weeks that followed, was that her deadline of four and one-half months for completing the study was impossible even if we had unlimited financial and personnel resources at our disposal.

During the next year, the four or us—three laboratory directors from Research Triangle Park in North Carolina, from Cincinnati, and from Las Vegas, and our chief in Washington—were fully engaged in finding toxic chemicals in that residential area and in explaining the significance of our findings. Many scientists on our staffs, at other EPA laboratories, and from private contractors supported this massive effort which left an indelible imprint on future approaches to environmental monitoring at waste sites.

In this area of about two miles by two miles surrounding Love Canal, the hunt for toxic chemicals was relentless. EPA specialists sampled the air near the Canal, between the houses, and along the roads. They removed samples of soil in yards, in fields, and in the schoolyard. To test the runoff from sewage and storms, they went into underground culverts in protective clothing. They scooped water from the river and streams and collected bottles of water from drinking water taps. The drilling of 175 sampling wells through yards and into shallow and deep groundwater was a particularly daunting experience. They also trapped and skinned field mice and took on the tasks of collecting earthworms and maple leaves which biologists advised were possible collectors of foreign chemicals.

Inside the homes they placed air samplers in bedrooms and living rooms. They probed into basement sumps. They positioned potatoes and oatmeal in places where food might absorb toxic fumes. They even shaved hair from pet dogs of the neighborhood in the search for accumulations of toxic metals.

Then EPA chemists shepherded many thousands of samples through analytical chemistry laboratories in distant parts of the country. Quality control was the byword. Every measurement was documented and cross-checked with other measurements. Questionable measurements were repeated or were discarded.

Meanwhile, other specialists collected supplementary information to provide a context for the chemical information. Hastily erected meteorological stations provided information on wind patterns in the area. Aerial photographs and topographic maps revealed depressions on the ground that could influence the movement of chemicals. Geophysical instruments sent radar signals, acoustic waves, and electromagnetic impulses into the ground to help us understand the flow of groundwater in the region.

Our instructions were very clear on one point. We were not to study the people: no blood samples, no urine specimens, no medical examinations. The local population had already been through such health surveillance several times. The earlier medical findings were repeatedly challenged as scientifically flawed and had led only to great confusion and controversy. The experts simply could not agree on the types of medical studies which would be useful in reaching a judgment on whether pollution had become a threat to public health. Therefore, the monitoring program was simply to characterize the chemical environment as the basis for an assessment of the habitability of the area. Of course, the study of the presence of pollutants was also to lead to conclusions about leakages of chemicals from the waste pile into the inhabited areas.

During my military service following World War II in Germany, I spent many days and nights in command centers surrounded by tanks, jeeps, artillery pieces, guards carrying gas masks, and barbed wire. At the Love Canal command center, I was surrounded by water tankers, dump trucks, bulldozers, drill rigs, police cars, fire engines, workers in moon suits, and high fences. In Germany, I had been concerned about potential foes behind the iron curtain. At Love Canal, we were facing very different adversaries, toxic pollutants and aggressive lawyers representing an irate public.

Returning to the design of the study, the initial task was to identify the chemicals to be placed on the detection "hit list." We reasoned that if we could determine the concentrations of the principal chemicals which were present in the area, then the toxicologists and the medical doctors could decide whether the levels were above or below thresholds of safety. But we had to decide in advance which chemicals would be the targets of the investigation so that our specialists could select the appropriate sampling and analysis equipment and techniques. For example, searching for a toxic metal which is as stable as a rock is different from probing for a chemical which turns into a gas when disturbed. Also, highly reactive acids need to be collected in different types of containers than inorganic salts which do not trigger similar reactions.

The process for selecting chemicals which were likely to be present in the area seemed obvious. First, available records identified dozens of chemicals which had been dumped by the Hooker Chemical Company into Love Canal many years earlier. Among those deposited in the largest volumes were the pesticide lindane and the industrial chemicals chlorobenzene, benzylchloride, sodium sulfhydrate, and dodecyl mercaptan. Second, the EPA had the results of earlier monitoring in the general area by New York state environmental specialists. Even though the thoroughness of these studies was being challenged, they nevertheless persuasively indicated the presence of other compounds and particularly toxic metals. Also, we had preliminary information on the chemical composition of liquids draining from the canal into a nearby collection system. Thus, we had a good starting point for preparing the initial hit list.

But as previously noted, the EPA was under siege by the Natural Resources Defense Council to search for the 125 priority pollutants throughout the nation's waterways. Some of these chemicals were already on the hit list. Many analytical laboratories which we had engaged for the effort were ready to analyze samples for all of the priority chemicals for a package price, and therefore we simply included all 125. Finally, as already noted, the EPA was seized with the problem of minute traces of dioxin throughout the country, and rumors were circulating that nuclear refuse had been placed in Love Canal at the end of World War II. Why not include these two categories of chemicals—dioxin and radioactive wastes—even though sampling for them required special handling procedures and the analyses required special laboratory techniques? So we included them as well in order to head off later criticism of inadequacies of our investigations.

We predicted that skeptics of the EPA's competence would still contend that the study had missed some chemicals. Indeed, within a few days after we received the assignment to carry out the study, staff members on Capitol Hill and government scientists in agencies other than the EPA began sniping at the approach we were developing. They argued, for example, that perhaps some chemicals placed in Love Canal many years earlier would decompose into other chemicals and these decomposition compounds would leak into the environment.

Therefore, we developed a last line of defense. We required each laboratory carrying out gas chromatography–mass spectrometry analyses, a technique which compared the peaks recorded on the computer screen as the samples were analyzed with peaks expected from the target chemicals, to also report unexpected peaks. These new peaks would indicate the presence of chemicals that had not been targeted in the study, EPA chemists correctly hypothesized. Thus, we would argue, no contaminants could slip through our net.

The sampling began in August 1980 and concluded by the end of October. We knew that such an extensive monitoring effort should never be conducted in such a rush. It should have been carried out in stages so that preliminary findings could guide more detailed investigations and so that sampling could take place in each of the seasons of the year as water runoff conditions vary. We truly needed one year on the site. But the political pressures for immediate results from the EPA Deputy Administrator and from the Congress were enormous. Therefore, after we had the concurrence of the state and local authorities and the reluctant cooperation of all local residents to go into the area, the EPA field teams simply had to move in and pull out as quickly as possible. Once the sampling was accomplished, the EPA scientists spent the next eight months analyzing the hundreds of thousands of bits of information which the chemical laboratories generated.

During this period, all the strengths and weaknesses of chemical-by-chemical analyses came to the fore. The program was highly successful in characterizing the area in general since the total number of chemicals which were investigated simultaneously was quite large, namely 150. On the other hand, considerable difficulty arose in explaining the origin and significance of local hot spots where a few chemicals were present and others which had been expected were not. Finally, the government did not equivocate in drawing conclusions as to the health implications of the findings as discussed below, but the basis for judging the potential problems associated with exposures to multiple pollutants at very low levels was a gray area for even the most sophisticated scientists.

In the end, the EPA concluded that the residential area of 600 homes was not being polluted as a result of Love Canal, and the Agency summarized the findings of the study as follows:

> . . . the (inhabited) area exhibited no clear evidence of Love Canal-related contamination . . . . Also, the data revealed that the occurrence and concentration levels of monitored substances . . . could not be attributed in a consistent fashion to the migration of contaminants from Love Canal. . . . Finally, the data suggested that the barrier drain system surrounding the landfill was operating effectively to intercept the lateral migration of contaminants from Love Canal. . . . [6]

These findings did not suggest that there were no man-made contaminants present in the inhabited areas. There were trace levels of many chemicals as is the case in all industrial areas of the United States. The findings showed rather conclusively, however, that the trace levels in the inhabited area could not be attributed to Love Canal.

Of course, the residents immediately wanted to know the public health significance of the contamination even if it was attributable only to general industrial pollution and not to the Canal. This same question had been asked before in other industrial communities, and health experts had always been reluctant to respond. The EPA had engaged a special contractor to conduct risk assessments of the findings of trace levels of the contaminants which were discovered near Love Canal, but the contractor gave up on the grounds that not enough was known as to how often the population came in contact with the chemicals.

The task of making a public pronouncement on the safety of the area based on the data we had collected fell to a senior government scientist from the Centers for Disease Control in Atlanta. Unlike many experts, he did not dodge the issue. His message was clear and unequivocal. There was no health risk from the chemical contamination.

The EPA had uncovered very low concentrations of several dozen contaminants (but not dioxin or any other equally potent chemical)—in the parts-per-billion and tens of parts-per-billion ranges. In the view of this medical expert, these levels were well below the threshold of concern. In public discussions he argued that the levels were consistent with levels found in other industrial cities around the country and there was no basis for believing that special health problems due to pollution existed in any of these areas. He later confided to us that only if the measurements had shown levels 100 times higher in the parts-per-million range would he have been concerned.

Still, some residents were not satisfied. The EPA had found a few samples in backyards which had contained one or two chemicals with concentrations 10 to 100 times higher than the average, and the residents wanted to know what these abnormal measurements meant for them. Also, in several areas we found low levels of eight or ten organic chemicals, and residents wanted to know the meaning for them when these measurements were aggregated. We could not provide satisfactory responses to their questions. In almost every environmental monitoring program in industrial areas, occasional high values of individual chemicals or aggregations of chemicals are discovered. The experts simply do not have good answers for questions that inevitably arise concerning an occasional exposure to these high values.

In general, the program was successful—scientifically and politically. The focus was on individual chemicals, but on so many that the comprehensiveness of the investigations could not be challenged. However, the program was unique, given the enormous effort expended by the EPA in carrying out the monitoring studies. Seldom will such high levels of scientific expertise and financial resources be available. The challenge is to adapt the approach of studying a "sufficient" number of individual chemicals for environmental assessments which are less well endowed.

Controlling the Most Toxic Chemicals with Concern for Many More

As we have noted, in a strict sense each chemical is different. It behaves in its own unique way. Each affects humans and ecological resources differently. Often, however, these differences are far more important to the scientist than to society. In some cases even the chemist has difficulty distinguishing one molecular structure from another let alone describing the differences in the environmental impacts of similar chemicals. While toxicologists and botanists have made impressive advances in determining the adverse effects of a few chemicals, they struggle with little success in predicting the synergistic effects of chemicals acting in unison on biological species.

Whether we are addressing toxic air or water pollutants, hazardous wastes, pesticides, industrial chemicals, or contaminants widely dispersed throughout the land, our nation's strategies for coping with environmental pollution must involve both control of individual chemicals and control of groups of chemicals. Some pollutants are simply too toxic not to receive special attention. At the same time, the number of potentially hazardous chemicals is too large to customize every regulatory approach to unique molecular structures.

Ideally, in their efforts to control groups of chemicals, regulatory agencies simultaneously constrain the individual pollutants of greatest concern. But this is not always the case. The level of control required for highly toxic chemicals may exceed the control levels imposed on other chemicals in the groups.

Also, placing limitations on the most toxic chemicals should address the most serious problems. But this also is not always true. The aggregated impacts of combinations of the less toxic chemicals have yet to be adequately explored.

The next chapter addresses the central issue in the control of chemicals—the magnitude and character of risks which they pose. However, the uncertainties associated with risk assessments are usually great. Consequently, the nation is often faced with a dilemma. Excessively stringent action to regulate individual chemicals may divert too many financial resources to environmental protection while inaction can lead to a more dangerous situation in the future. Controlling groups of chemicals by requiring industry to use the best available control technologies without trying to define the precise level of risk, as is now required under the Clean Air Act, is one concrete example of how this dilemma might be addressed in the future.

At the same time, governmental overreaction to health threats from individual chemicals needs to be tempered. For example, the tens of millions of dollars currently being spent to rip asbestos insulation out of schools and public buildings throughout the country exemplify how well-intended efforts have gone out of control simply because it was easy to peg "corrective" actions to a specific chemical. Asbestos can be a serious problem when it breaks up and enters the air; but as long as it is embedded in solid building materials, the best strategy is to let it remain for the lifetime of the material and then arrange for careful disposal.[7]

On the other hand, a responsible approach focused on a single chemical has been adopted by the Congress in controlling lead in drinking water, with particular attention to exposures of children and pregnant women. Specifically, legislation requires repair or removal of water coolers with lead-lined tanks and a ban on future sales of such coolers. There are alternative products. Also, the legislation provides funds for medical screening of children and calls for remedial action when problems arise involving lead in plumbing and in kitchen facilities.[8]

3 The Uncertainty of Risk but the Reality of Cost

Security is mostly a superstition. It does not exist in nature, nor do the children of men as a whole experience it. Avoidance of danger is no safer in the long run than exposure. Life is either a daring adventure or nothing.
—Helen Keller

Uncertainty, in the presence of vivid hopes and fears, is painful, but must be endured if we wish to live without the support of comforting fairy tales.
—Bertrand Russell

The concept of the mythical man was introduced in the debates on the Clean Air Act—stand by a plant's fence for 70 years, breathe the chemical 24 hours a day, and have a one in one million chance of getting cancer. If that is the case, then shut down the plant. These theories are lost on my colleagues and myself.
—Senate Minority Leader Robert Dole

Toxic Trouble in the Plastics Industry: Vinyl Chloride

"How do you explain the behavior of those geriatric rats?" asked the EPA Administrator. He wanted his medical advisers to explain why three laboratory rats which developed cancerous tumors after being dosed with the toxic chemical vinyl chloride (commonly called VC) had outlived 47 dosed rats which had not developed tumors and 50 rats in a control group which were not exposed to the chemical at all. The life expectancy of rats is 104 weeks, and 97 rats died on schedule. But the three which were the center of attention had survived until they were 110 to 120 weeks old. "Didn't the three tumor-laden rats simply die of old age?" queried the Administrator.

The experts surmised that the tumors probably developed very slowly and therefore afflicted only the oldest rats. They vigorously argued that despite the uncertainty as to how the tumors evolved, the tumors were real. They were of a unique type known as angiosarcoma of the liver, noted the experts. Therefore, they could be unequivocably linked to the exposure of the rats to VC. Since several workers had died from exposure to VC, no one doubted this linkage. What was on the table for discussion was the *level* of exposure to VC that could cause cancer, and the experts were presenting this laboratory experiment as an important development in determining a hazardous level of exposure.

The Administrator was not belittling the importance of research findings. He simply was trying to understand the relationships between the impact of toxic agents on laboratory animals and their effects on human beings. The Administrator, like his predecessors and successors, was facing the dilemma of weighing limited scientific information in an effort to balance human lives against the expenditures of large national resources required to maintain the viability of probably the most important segment of the American plastics industry. The issue was not *whether* regulation was needed. Rather, the question was *how much* regulation was appropriate. This will continue to be the common problem in controlling toxic chemicals.

An unexpected event led to the May 1974 gathering of the EPA Administrator and his advisers which turned out to be one of the most significant meetings of environmental officials during the mid-1970s. For the first time the toxicity of a chemical threatened the future of an entire industry with annual sales in the billions of dollars.

In January of that year, the B.F. Goodrich Company, the largest American producer of the plastic polyvinyl chloride (known as PVC) which is widely used in consumer and industrial products, notified several government agencies including the EPA that four workers from its PVC manufacturing plant in Louisville, Kentucky, had died from angiosarcoma of the liver. The Department of Labor required such notification of job-related injuries or deaths. The Goodrich officials reported that these four employees had worked for many years in areas of the plant where significant amounts of VC pervaded the air. Goodrich therefore concluded that exposure to VC had caused the cancers. The EPA Administrator thereupon asked me, as Director of the Agency's Office of Toxic Substances, to lead the EPA's effort in responding to this rather frightening development.

This plastics plant and about 35 other manufacturing plants throughout the country used VC as their initial raw material in the production of PVC. Engineers injected the VC gas into large sealed kettles where a series of chemical reactions at high temperatures and pressures polymerized, or transformed, the gaseous VC into PVC which is a solid material. At that time, the workers in many plants handled the VC raw material in a very sloppy manner despite a general awareness throughout the industry that it had toxic characteristics. Leakages of VC at loading docks and through old valves and fittings within the plants were commonplace. In addition, the reaction processes in the kettles were not well controlled, and large amounts of the VC were not converted to PVC. Residual VC remained either as a waste gas in the kettles or as gas entrapped within the PVC plastic. Subsequently, during cleaning of the kettles or processing of the plastic, VC would leak into the workplace or into the general environment.

The rarity of that particular cancer—angiosarcoma of the liver—and the clustering of four deaths at the Goodrich plant raised immediate flags within the chemical industry and the government that a very serious occupational hazard had been uncovered. Soon government investigators traced ten more deaths of former workers at other plants to VC exposure, and the press reported two cases of the liver disease among nonworkers who had lived near a PVC plant.

The EPA's most immediate task was to determine whether the several million Americans living within a few miles of PVC manufacturing plants were at risk from VC discharges and, if so, to take steps to reduce any such risks. Thus, the first priority was to investigate the extent of the escape of VC from the 35 PVC plants. However, my EPA colleagues and I also suspected problems at some of the 15 other plants throughout the country which produced the VC raw material. In addition, 7500 plants made products from PVC which might be impregnated with residual VC that could leak out as the plastic was molded into different forms. As another concern, plastic pipes made from PVC had become very popular within the home building industry for bringing drinking water into residential areas, and in theory VC temporarily trapped in new pipes might end up in tap water. Finally, industry used VC to improve the spray properties of several aerosol propellants and also to enhance the homogeneity of a few specialty coatings used to help preserve industrial and consumer products.

In order to address each of these concerns, EPA specialists needed to develop correlations between the levels of exposure of humans to VC and the likelihood of harmful effects. From laboratory studies, they knew that VC caused liver cancer. However, they did not know what level of exposure would trigger the onset of cancer, and they did not know what other adverse health effects could be linked to VC. Operating on a very short time schedule, they had to make do with available scientific data which were far from satisfactory. The commissioning of additional studies with laboratory animals would have been extremely useful. But such studies would have taken six months to three years to carry out, and the pressure was on for immediate decisions.

The EPA Administrator obviously wanted a regulatory approach for dealing with the VC problem very promptly—an approach which in the first instance responded to the public health threat but which also took into account the economic importance of the PVC industry to the nation. One of the choices available for the EPA was to shut down the worst offending plants immediately. Or the Agency could start the process of establishing a national standard to limit air emissions of VC, a procedure which might drag on for two to three years to permit all interested parties to voice their concerns. Of course, the EPA needed to consult with the Occupational Safety and Health Administration of the Department of Labor concerning steps that agency might take to protect workers—steps that would also reduce VC emissions into the general environment. Finally, the EPA needed to coordinate efforts with the Food and Drug Administration and other agencies in addressing plastic pipes in drinking water systems since at that time the EPA's legal responsibility for regulating pipes was uncertain.

The economic stakes were very large. PVC was a widely used plastic and was a backbone of many branches of industrial and commercial activity in the apparel, building, construction, home packaging, recreation, and transportation sectors. The annual wholesale value of PVC products in the United States was billions of dollars. Involved in VC production were 1500 workers; 5000 more were engaged in PVC production, and 350,000 in molding finished plastic products from PVC.

The industry was growing at a rate of 14% annually. Some American companies had effectively penetrated foreign markets, and the industry had established a clear leadership role internationally. Still, firms in other countries were attempting to challenge this leadership. If PVC production costs increased in the United States as a result of environmental controls, imports of plastic from Europe would become far more attractive to American wholesalers at a time when balance of trade was an important political issue in Washington.

During the next four months the EPA mobilized its offices and laboratories nationwide in an aggressive effort to clarify the harmful effects. "What steps could industry take immediately to help correct the problem with a minimum of financial loss?" we repeatedly asked.

Most manufacturers of PVC were quick to acknowledge that while they had not violated any laws or regulations in producing the plastic, they had been lax in not exercising greater care to contain VC. Also, industry itself had taken the initiative several years earlier to sponsor laboratory studies which pointed to the carcinogenic properties of VC in rats. But the sponsoring companies had delayed for many months reporting the results of these studies to the broader scientific community or to the government until Goodrich revealed the deaths of its workers. Industrial representatives reluctantly admitted that this sheltering of scientific data had been a serious mistake even though at that time there was no legal requirement to release the results of animal studies that were voluntarily sponsored. When the issue of corrective action to reduce VC emissions was raised, industry spokesmen pleaded for patience. They argued that they could not make radical engineering adjustments without incurring enormous costs that would result in the closing of some plants.

Returning to the decision meeting with the EPA Administrator, the EPA staff entered the conference room armed with large notebooks of facts and allegations. Their voluminous documents included pollutant measurements around facilities throughout the country, assessments of manufacturing processes within the plants, reviews of epidemiological and laboratory studies of the health effects of VC, and reports of consultations with hundreds of scientists and engineers.

About 20 of the EPA's best specialists characterized the problem for the Administrator and recommended solutions. They tried to reconstruct the levels of exposure which led to the deaths of some workers as well as the levels which apparently had no effect on thousands of other workers. These studies indicated that the harmful levels were in the range of several hundred parts per million (ppm). Analyses of the toxicological studies depended on extrapolating the effects on humans from animal experiments. These studies suggested much lower levels of concern in the range of 1 to 50 ppm.

Why was there such a large disparity? The toxicologists were very conservative in extrapolating effects on animals to effects on humans on the one hand, while other health experts were quite pragmatic in simply relating observations of the presence or absence of tumors in humans to their estimate, albeit very uncertain, of the levels of exposure. In any event, all of us were concerned by the measurements on the front porch of a home adjacent to the Goodrich plant in Louisville of 40 ppm and by many other household measurements above 1 ppm near several plants.

As to the economic costs of regulatory action, the EPA experts contended that by tightening valves and fittings and by simply improving waste handling procedures within the plants, plant managers could easily reduce the levels of VC emissions. They also concluded that by paying greater attention to operating temperatures and pressures, shift supervisors could reduce the quantities of residual VC gas that remained in the kettles. But the representatives of some companies whom the EPA staff had queried disagreed, particularly those from companies with older plants. They had argued that significant reductions of VC emissions depended on installing entirely new production lines—an undertaking which they could not accomplish overnight.

Having listened intently to the presentations, the EPA Administrator reported briefly on his discussions with White House staff members, senior officials of the Department of Labor, and the head of the Food and Drug Administration. He noted that they were as anxious as the EPA staff to learn his decision, but they had provided little additional information which would influence that decision. They simply stated that the responsibility rested with the EPA and that they would try to ensure that their future steps complemented the EPA's actions. He was particularly disappointed that the Department of Labor had not moved more aggressively to take corrective action and reduce worker exposures. That Department was well equipped to prevent accidents in the workplace, but it simply did not have the scientific or engineering capability to deal with the complexities of worker exposures to low levels of chemicals.

The Administrator then decided on several courses of action. First, he would summon to the EPA headquarters in Washington the chief executive officers of the 25 companies operating VC and PVC plants and inform them that they had 30 days to reduce the VC emissions in each of the plants. This corrective action was to be "voluntary" on their part, but with the clear implication that the EPA would take legal action to close some plants immediately thereafter if the levels were not reduced significantly. The Administrator correctly surmised that when confronted with this edict, industrial managers would find under-utilized technical means to help correct the problem. Second, the EPA would begin the process of establishing an air emission standard for VC: discharges from plants could not result in a concentration exceeding 1 ppm at the fence lines of the plants, according to the proposed standard. Next the EPA would urge the Department of Labor to reach out to the scientific community for assistance in developing the technical justification for immediate steps requiring industry to reduce levels of VC exposures of workers. These steps would also curtail emissions of VC into the general environment. Finally, the EPA would continue its research and would investigate actions to further clarify all aspects of the VC problem.

Within a few days 25 company limousines encircled EPA headquarters as top industrial officials from across the country assembled to learn of these decisions. Managers of both old and new plants unanimously agreed that they could indeed reduce emissions significantly through engineering adjustments, and they did. The levels of VC came down, and no plants closed for environmental reasons.

Meanwhile, the EPA's studies of drinking water pipe and of the molding of plastic products indicated that very minute quantities of VC occasionally leaked from the plastic material, but the levels were of little concern. Also, as the manufacturing processes for converting VC to PVC were tightened to reduce losses of VC which should have been converted to PVC, the problem of residual VC impregnating the plastic would disappear almost entirely.

Compared to many other environmental rulings, the Administrator's decision concerning the production of VC and PVC was relatively easy. No one doubted the seriousness or the cause of the problem.

Indeed, VC is almost unique among environmental chemicals in triggering a disease that can only be attributed to a single chemical. Also, the potential economic consequences were sufficiently large that American engineering ingenuity would be extended to its limits. Finally, the scientific data base, including the data about the geriatric rats, was better than usual.

The decision clearly illustrated many facets of assessing chemical risks and taking immediate steps to reduce risks. First, all available information on health effects must be considered, including both data from carefully designed toxicological studies and information about people who have been exposed to the chemicals of concern. Inevitably wide bands of uncertainty surround such studies and information, but decisions must be made in spite of this uncertainty. Second, programs for the monitoring of concentrations of toxic chemicals in the air, water, food, and other materials during the times of environmental crises and also as a normal course of everyday business are especially critical in determining the extent of risks. There simply is no substitute for authoritative sampling measurements of chemicals suspected of threatening the health of humans.

Also, in responding to chemical risks, government agencies should recognize that industry can usually offer better technical solutions to solve specific problems than can the government. Frequently a push from the government is necessary. But once energized, industrial ingenuity is usually much more effective than technical solutions developed by government experts.[1]

A Threat to the Health of Newborn Infants: PCBs

One year later, I became involved in another aspect of balancing risks and costs. A different toxic chemical posed a risk, and this time a different type of cost was at stake. Exposure to this chemical might cause learning disabilities in children, a loss to families and indeed to society at large.

In mid-1975 the EPA discovered that nearly all nursing mothers in the United States were feeding breast milk contaminated with PCBs to their newborn infants. For a number of years the EPA had routinely collected milk samples from about 1500 mothers at hospitals in different parts of the country and analyzed these samples for the presence of trace levels of pesticides. In the early 1970s, as the EPA's laboratory analysis methods became increasingly sensitive in detecting low levels of contaminants, government chemists began discovering the presence of trace levels of PCBs in the milk samples in addition to several pesticides. Thus, the revelations in 1975 were not completely unexpected.

However, the 1975 results suggested that the problem was more serious than previously recognized. The alarming aspect of the new reports of PCBs in mother's milk was the high levels found—levels of 1.0 to 5.0 ppm. The EPA's Office of Toxic Substances again became the focal point within the Agency for dealing with an emotionally charged problem.

At that time, a research scientist at the University of Wisconsin had been investigating the effect of PCBs on young monkeys. He fed PCBs to female parent monkeys who then passed the chemical on to their infants through breast feeding. He tested the breast milk and watched the behavior of the newborn monkeys for several years. He reported to the scientific community that the levels of PCBs in the breast milk were in the range of several parts per million. According to him, observations showed learning disabilities among the offspring.

By 1975, PCBs had become ubiquitous throughout the environment. The EPA had found the chemical in fish, in sediments, and in soil in many regions of the country. Very low levels were even showing up in drinking water. Of course, from the EPA's program of sampling mother's milk, Agency specialists had known that PCBs were accumulating in people. Also, the EPA had a special program for testing tissues from human corpses nationwide. These tests identified the presence of PCBs, also at levels of several parts per million.

EPA specialists knew of no way to control such widespread contamination in the short run. The manufacturers of PCBs had stopped production. PCBs still in use were being contained in a more responsible manner than in the past. Leakages of PCBs into the environment from waste piles and from contaminated products were gradually being cleaned up or contained. But the impact of these actions on the environmental presence of PCBs would not be seen for some years.

In addition to immediate concerns over the health of newborn infants, we knew that we in Washington would have a massive public relations problem on our hands once the press learned about PCB contamination of mother's milk. What advice should the government give to nursing mothers? The EPA turned to the Department of Health, Education, and Welfare (now reconstituted as the Department of Health and Human Services) and to its Public Health Service to provide the answer to that question.

The senior officials of the Department were not eager to take on this issue. The Department was still reeling from the aftermath of the advice they had given to the American public the previous year to undergo inoculations for warding off swine flu. This advice had resulted in a barrage of reports of adverse side effects from the inoculations. Thus, the Department's specialists were hesitant to take a stand concerning PCB contamination. Still, they too recognized that it was only a matter of time until the press would have the story, and they would be expected to say something.

The immediate issue was whether the Public Health Service should develop a nationwide Health Advisory bulletin concerning contamination of mother's milk. Should they advise the medical community not to encourage breast feeding or alternatively should they alert doctors of the need to consider the pros and cons of breast feeding in the light of this new information? Even if a health advisory were not issued, what position would be taken in response to the inevitable press inquiries? The EPA operated in a glasshouse environment and simply could not prevent the data from reaching the public domain very quickly.

Meanwhile, EPA scientists were raising questions concerning the validity of the monkey studies. One report reaching the government indicated that the monkeys were not treated properly and had become sick during the experiments. Therefore, attributing their behavior to PCBs was questionable. Other reports suggested that the methods for administering and measuring the dose levels of PCBs were faulty.

In any event, after a few days of handwringing the EPA agreed with the Department's plan to deal with the problem. The Department would convene a *public* meeting at the National Institutes of Health on the edge of Washington where a panel of eminent medical experts would discuss the problem. Under the watchful eye of press representatives, they would recommend actions that should be taken by the Department. The experts would include both health practitioners and research toxicologists.

My role at the meeting was to lead the presentation of the EPA's findings of PCBs in mother's milk and to orchestrate the discussion of EPA's assessment of the extent of PCB contamination in air, water, fish, and soil throughout the country. Then I was to take a back seat and let the Department's representative direct the rest of the meeting devoted to the health implications.

The medical experts at the meeting included six obstetricians, several general practitioners from the mid-West who had delivered and cared for literally thousands of babies, and three toxicologists. Two prominent observers were a representative of the La Leche League, an organization dedicated to the promotion of breast feeding, and a representative of the canned milk industry. A handful of journalists sat in the back row.

The general practitioners vigorously defended the many health benefits of breast feeding. They stressed over and over both the direct physiological benefits and the more subtle psychological benefits of nursing. The obstetricians supported these views. The toxicologists described the monkey experiments. They were hesitant to comment on the validity of the studies and added little concerning the significance of these experiments for the breast feeding of humans. They offered few insights as to the seriousness of this type of contamination which was being addressed for the first time. The La Leche League representative was quite effective in her summary of testimonials supporting breast feeding by many mothers. In contrast, the industry representative correctly decided that silence was his best course of action since the enormity of the health considerations clearly dwarfed business factors which would have been ascribed to his viewpoint.

The summary of the meeting by the Department's chairman stressed the virtues of breast feeding and the need to conduct more research on the effects of contaminants in mother's milk. He emphasized the recommendations for intensive follow-up studies of the learning capabilities of a sample of children known to have been fed contaminated milk.

The idea of a health advisory was smothered before it was even proposed. The general practitioners from the mid-West persuasively argued that the government should keep quiet. They contended that any statement from Washington would simply confuse doctors and mothers throughout the country who should not be deterred from breast feeding.

The next day I received a telephone call from a colleague in the

Public Health Service who had played a major role in organizing the meeting. He opined that the meeting had been a great success. He pointed to the consensus for a continuation of breast feeding that developed around the table. He added that the Department was especially relieved that the *Washington Post* had not reported the meeting on the front page and had not called for a health advisory.

This problem again pointed out the wide range of uncertainties in dealing with the health effects of toxic chemicals, uncertainties which were compounded by the questions raised over the validity of the principal scientific study being considered. Also, it highlighted some of the types of trade-offs that are encountered in weighing the merits of actions to limit specific risks. However, placing the responsibility for decision making in the hands of an ad hoc panel of scientific experts, even though the experts were carefully chosen by the government to consider both sides of the issue, was a little unusual.

Finally, as is often the case, the press was a major factor in forcing the government to reach a prompt decision on how to respond to a chemical problem regardless of uncertainties in the assessments of risk.[2]

Quantifying the Risks from Toxic Chemicals

During the 1970s, the EPA and other regulatory agencies directed much of their attention to the immediate problems posed by VC, PCBs, and a few other highly publicized toxic chemicals, and particularly to the uncertainties in dealing with their risks to public health. As a consequence of the difficulty in coping with these uncertainties on a crash basis, the need for scientifically credible methodologies for assessing health risks dominated many interagency discussions. Within the EPA, in particular, several offices strengthened their scientific capabilities to assess risks soon after the VC episode.

Meanwhile, economists at the White House Office of Management and Budget and elsewhere within the government pressed for detailed studies of the costs to the economy—both direct and indirect—when regulating chemicals. They insisted that the EPA attempt to balance these costs against health risks in reaching regulatory decisions. In order to carry out such a balance, argued the economists, the risks and costs needed to be quantified. Soon the "art" of quantitative risk assessment began to emerge.

The EPA thereupon devoted considerable manpower to quantifying the risks from chemical exposures to carcinogens since the concern over cancer was a driving force in seeking new regulatory authorities. Typically, toxicologists administered the suspected carcinogen to laboratory rats or mice at different dose levels, including the highest possible level which would not cause immediate death from poisoning. The number of animals that eventually developed tumors at each dose level could be determined during autopsies, and the probability of a single animal developing tumors was calculated. The biostatisticians then plotted on a graph in the form of a continuous curve the probabilities that the different dosages would cause cancer. Thus, the scientists quantified the carcinogenic risk of a chemical—at least the risk to laboratory animals at high dose levels.

The next step was to relate these findings from the studies of animals subjected to high doses of chemicals in the laboratory to the effects on humans at the much lower pollutant concentrations commonly encountered in the environment. The statisticians extended the curve downward into the lower ranges of exposure. They relied on the scientific judgments of toxicologists as to the biological activity of the chemical at these lower levels since experimental data to guide the extension of the curve were not available. Interminable debates continue to this day, however, over the shape of these curves as the chemical concentrations decline down to the trace levels of contamination in the environment.

Controversy abounds over the way different chemicals behave in the bodies of both laboratory animals and human beings at high and low dose levels. Do they trigger formation of other chemicals? How do they promote tumors? Do some chemicals inhibit the development of tumors? Why do some chemicals cause tumors in one animal species and not in another? Despite these and many other uncertainties, scientists are required by the regulatory agencies to provide their best estimates of the likelihood that cancer will develop from low levels of exposure to chemicals which are identified as carcinogens, and they render judgments.[3]

The substantial differences between the physiological characteristics of humans and those of rodents undoubtedly influence the onset of cancer. When providing risk estimates for regulatory agencies, toxicologists assume that humans are much more vulnerable than are experimental animals to cancer induction from exposures to chemicals. Therefore, they incorporate safety factors into the process of extrapolating from animals to humans to take account of this increased sensitivity of humans. The toxicologists then present the data about risks to humans in quantified probabilities.

In 1973 the EPA published in the *Federal Register* its first formal quantitative estimate of the risk of cancer posed by human exposure to a toxic substance. The chemical was benzidine—a chemical which was widely used at that time in dyes.[4] Since then, the Agency and many other governmental and nongovernmental bodies have carried out hundreds of quantitative risk assessments which have been used as the basis for regulating carcinogens.

In responding to the economic concerns of the Office of Management and Budget, the EPA also strengthened its economic assessment capabilities. Soon the dialogues over risk within the Agency took new tacks. For example, the advocates of cost–benefit analyses argued about the likelihood of cancer developing in a specific individual as the result of exposure to a specified level of a carcinogen. Was it one chance in a thousand, one in a million, or one in ten million? Knowing the answer, the decision maker could then balance costs and risks and attempt to determine the price of life that would be associated with specific regulatory actions, continued the logic of the argument.

One type of analysis proceeded as follows. If the public health risk of not taking a regulatory action to control a carcinogen could be quantified as the likelihood of 100 cancer deaths nationwide and if the cost to the economy of regulation could be estimated at $10,000,000, the responsible regulatory official could decide whether life should be valued at more or less than $100,000 per person. His or her value judgment would then drive the decision to regulate or not regulate.

The remnants of this type of simplistic analysis now reside largely in think tanks and academic settings. Few decision officials are prepared to consider the evidence in such narrow terms. Few members of the public will agree to put a price tag on the lives of loved ones. Nevertheless, the price of life remains a lively issue in the courtroom hearings on toxic litigation directed to compensation for alleged injuries from environmental contaminants.

Another type of analysis estimates the costs to society in caring for the victims of pollution. For example, one high estimate by a health advocacy group is that the adverse health effects of air pollution from automobile exhausts currently cost the nation about $50 billion annually. This estimate includes both lost productivity due to absences from work and the costs of providing health care for those persons who are particularly sensitive to such pollution. Such estimates buttress the case for stronger controls. The industries which must abide with air pollution laws point out the costs to the nation of compliance with new controls—in this case the industry presents its high estimate of about $40 billion per year. This would seem like a reasonable trade-off, but political complications arise since those who most bear the cost are not necessarily those who are suffering the adverse impacts.[5]

Clearly, quantitative risk assessments can be helpful, and in many cases essential, in clarifying the toxicity traits of chemicals and the severity of problems they can cause. Similarly, quantitative estimates of economic impact are important to provide a perspective for responsible decision making. But making direct quantitative trade-offs between public health and economic impacts as the sole basis for reaching decisions on regulating chemicals is unrealistic. Too many scientific and economic uncertainties and too many nonquantifiable political and social factors are involved in such decisions. In the words of a report of the National Academy of Sciences: "Benefit–cost analysis should be thought of as a set of information-gathering and organizing tools that can be used to support decision-making rather than as a decision-making mechanism itself."[6]

During the past decade, quantitative risk assessments of chemicals believed to be carcinogens have continued to be a major activity for many regulatory agencies in Washington and throughout the country. Highly sophisticated computer models predict the relative carcinogenic potency of many chemicals, and other models provide the basis for estimating the likely exposures of populations to chemicals. Extensive guidelines for carrying out quantitative risk assessments have been developed and endorsed by the nation's leading scientists, within and outside government, and even the courts have demonstrated a remarkable degree of familiarity with and dependence on assessments based on these guidelines.[7]

In taking steps to codify their approaches to quantifying carcinogenic risks, the federal regulatory agencies point out that "three types of evidence can be used to identify substances that may pose a carcinogenic hazard. . . (1) epidemiologic evidence derived from studies of exposed human populations, (2) experimental evidence derived from long-term laboratory studies of animals, and (3) supportive or suggestive evidence derived from studies of chemical structure or from short-term or other tests that are known to correlate with carcinogenic activity."[8]

The most persuasive evidence that a chemical induces cancer is an epidemiological study which shows a high correlation between groups of individuals exposed to the chemical, such as industrial workers, and unusual numbers of cancers within the groups as was the case with VC. The number of chemicals which now can be clearly labeled as "human" carcinogens based on such studies is about two dozen. However, epidemiologists have hesitated to develop techniques for providing quantitative estimates of risk from these chemicals, usually arguing that *any* exposure to a known human carcinogen is unacceptable.

Laboratory studies which show that chemicals will induce cancer in experimental animals are much more common and are the principal determinants in classifying chemicals as carcinogens. As discussed, they also provide the data for quantitative risk assessments. The government has indicted about 200 chemicals as carcinogens based on evidence from long-term animal studies. Many more chemicals have shown carcinogenic tendencies in less comprehensive animal studies or in other types of laboratory investigations.

The regulatory agencies repeatedly acknowledge the difficulties in extrapolating from high doses in the laboratory to low doses in the environment and from experimental animals to humans. In some cases the estimate of potency can vary by more than 10,000 times depending on the extrapolation model which is chosen.

Central to risk estimation within the regulatory agencies is the following philosophy: ". . . current methodologies which permit only crude estimates of human risk are designed to avoid understatement of the risk."[8] In short, the agencies are conservative and err on the side of safety. In this regard, scientists are not able to prove that carcinogens will have no health impact at even the very lowest conceivable level of exposure. In the words of the regulatory agencies, ". . . exposure to any amount of a single carcinogen, however small, is regarded as capable of adding to the total carcinogenic risk."[8] Such statements are repeatedly used by environmental extremists in advocating bans on all carcinogens regardless of economic considerations.

In addition to the toxicity of a chemical, the extent of exposure of individuals to the chemical is a critical factor in determining risks. At one end of the spectrum, if the chemical is produced exclusively in an industrial process and survives only for a millisecond inside a sealed reaction vessel until it is changed into another form, then the chemical is of no risk to society regardless of its toxicity. At the other extreme, workers who breathed asbestos on the job in shipyards for many years obviously were at very high risk.

Determining the extent of human exposure to chemicals is even more difficult than estimating the potency of the chemicals. Many forms of cancer and other chronic diseases develop slowly during a period of 20 years or more. Thus, histories of human exposure patterns over a long period of time are important. But relating the life-style of a single individual, let alone a group of individuals, to the presence of a chemical over a period of years is fraught with uncertainty. Then relating human exposures to exposure of laboratory animals can be even more speculative. For example, humans may be exposed to a chemical for a few seconds or minutes or intermittently over a period of many years whereas laboratory animals are often continuously exposed to the chemical for their entire lifetimes.

Should risk estimates be oriented toward exposures of specific individuals to chemicals or to exposures of large groups of people? Obviously both are important. However, it is often easier to present risk estimates in terms of risks to population groups than to specific individuals. An illustrative statement might be: If 10,000 people are exposed to an air pollutant for five years at a specified concentration, eight additional cases of cancer will develop. A much different type of statement is the following: If my sister who is in her early thirties and is a heavy smoker lives among that population group, her chances of having lung cancer by age 50 are increased by 1%. Most risk assessment methods are oriented toward the first example of considering aggregated populations.

Despite the difficulties in estimating the potency of toxic chemicals and the still greater difficulties in reconstructing or predicting exposures over a long period of time, decisions must be based on both of these considerations. If ever there were areas of science crying for more intensive research, chemical toxicity and exposure assessment are the areas. While science will never provide definitive answers, research can certainly provide better tools for improving the basis for judgments, and particularly research aimed at evaluating laboratory experiments with animals in light of epidemiological investigations of human populations.

Finally, we as a society need to accept the concept of *de minimis* risk of toxic chemicals—a level of risk that is so low that it should be of only minimal concern to regulatory agencies. This concept has been used for many years in the field of nuclear radiation. Normal exposure to sunlight and x rays, for example, is commonly accepted, and costly regulations to limit exposure to still lower levels of other types of radiation should be accorded very low priority.

Similarly, there is risk from exposure to naturally occurring toxics in general and carcinogens in particular which we must tolerate in our daily lives. As in the case of radiation, society should accept minimal levels of man-made chemical exposures as the price for living in an industrial society and should concentrate on preventing the higher levels which are much more harmful.

The most commonly suggested *de minimis* risk level for carcinogens is a level of exposure that will induce cancer in one person in a population of 1 million persons who are exposed to the chemical. This is a small risk indeed in comparison with the overall rate of 230,000 of every 1 million deaths in the United States being attributable to cancer from all causes. Of course, the government should encourage steps to minimize exposures to all types of chemicals but should press for regulation of those that make a significant difference. Government officials should not fuel the notion that zero risk of cancer from man-made chemicals is an achievable goal, either technically or economically.

The foregoing discussion highlights the enormous effort which environmental specialists throughout the country have devoted during the past 15 years to understanding and quantifying the risks posed by carcinogens. Their research efforts have helped clarify the magnitude of risks and have repeatedly pointed out the uncertainties in developing risk estimates. This valuable research should continue apace.

At the same time, there has been an undesirable consequence of quantitative risk assessments. They have been targeted almost exclusively on carcinogens, inadvertently diverting attention from many other types of risks of comparable importance. Some chemicals can induce heart disease, birth defects, genetic changes, and nervous disorders, for example. Since such impacts of chemical exposures are not easy to detect (i.e., they do not have signatures such as tumors which can be easily recognized), scientists have great difficulty persuading the public of their importance and their possible linkages to chemical exposures. Also, scientists are at a loss as to how to quantify these adverse effects. Thus, the public now has an exaggerated view of the *relative* threat posed by man-made chemicals which have carcinogenic tendencies.

More than a decade ago, a leading American scientist foresaw this distortion in the public's perception of the carcinogenic hazards of man-made chemicals and pointed out: "Much of the cancer occurring today, in addition to that caused by cigarette smoke and radiation (such as ultraviolet light which induces skin cancer), appears likely to be due to the ingestion of natural carcinogens in our diet. For example, fat intake has been correlated with breast and colon cancer, and many plants used in the human diet have developed a wide assortment of toxic chemicals (probably to discourage insects and other pests from eating them) that may be mutagens and carcinogens. In addition, powerful nitrosamines and nitrosamide carcinogens are formed from certain normal dietary biochemicals containing nitrogen, by reaction with nitrite. Nitrite is produced by bacteria in the body from nitrates that are present in ingested plant material and water. A number of molds produce powerful carcinogens such as aflatoxin and sterigmatocystin; these molds can be present in small amounts in foods such as peanut butter and corn."[9]

The important point is that the cancer risk from man-made chemicals should be kept in perspective. We are frequently exposed to higher levels of natural carcinogens than man-made carcinogens. Plants produce carcinogens for their own protection against insects, and thus some vegetables and grains contain carcinogens. Also, the overall cancer rate for the United States has remained essentially constant for the past 50 years even though the production of chemicals has increased dramatically. Of course this does not mean that we should not try to limit exposures to man-made carcinogens, for we should; and quantitative risk assessments are critical in determining strategies for limiting exposures and reducing risks.

On Being Exposed to Mixtures of Chemicals

For the past two decades, most of the national effort in Washington, in state capitals, and in scientific laboratories to improve the scientific basis for regulating toxic chemicals has been based on studies of individual chemicals. Yet, as we have seen, some of the most pressing toxic problems result from exposures of people or ecological resources to many chemicals at the same time.

I have joined teams of EPA specialists investigating fishkills in polluted streams in Montana, Texas, and New York. These streams were threatened not by single pollutants but by multiple contaminants. In Montana, several toxic metals were draining from mining areas. In Texas, factories refurbishing aircraft were flushing both metals and organic solvents into the stream. In upstate New York, many toxic organic chemicals from a poorly operated sewage treatment plant had destroyed all aquatic activity.

We have all stood in the wake of heavy-duty trucks spewing diesel exhausts. When we drink heavily chlorinated water, we hope that the strange taste is a signal that good chemicals override both bacteria and bad chemicals. And when we see those barrels of toxic wastes, we urge that they be taken far away lest they corrode and begin to leak near our homes.

Further, our scientists must now expand their limited efforts of the past to assess the likely problems associated with mixtures of chemicals. While toxicologists may be able to ensure that their experiments expose laboratory animals to chemicals one at a time, real-life experiences are surrounded by chemical mixtures in our industrial cities. Mixtures of chemicals lurk in living rooms bedecked with plastic and subjected to a variety of household cleaning agents, and in restaurants mixtures of artificial food ingredients are often considered a key to elegant dining. Some doctors are increasingly asking, "Can we attribute many allergic reactions—sneezing, hives, and rashes—to our constant exposure to seemingly insignificant levels of many man-made chemicals?"

In a few cases, industry and government laboratories routinely carry out studies on mixtures. One example is the testing of pesticides. Pesticide chemicals are seldom sprayed in the field in pure form. The materials applied to crop areas, for example, usually contain small amounts of chemicals which help in the spreading of the pesticide. Also, very small amounts of waste by-products from the processes of manufacturing pesticides may be present. Consequently, regulations for approval of pesticides require testing of the "technical-grade" materials—the materials used in the field which are actually mixtures of chemicals.

Often, scientists attempt to assess the properties of newly encountered mixtures based on tests that have been conducted on similar but not identical mixtures. The new mixtures may contain the same individual chemicals in different ratios or may include most but not all of the chemicals. For example, studies of the impact of a wide variety of diesel exhausts on laboratory animals have been conducted for years. Even though diesel exhausts may vary from vehicle to vehicle, generalized inferences concerning the health effects of new mixtures of diesel fuel are made from the extensive data base already developed.

More commonly, however, specialists attempt to assess the risks associated with mixtures using studies of the individual chemicals which make up the mixture. If two or three chemicals dominate the mixture, the task is simpler than if there are a dozen chemicals present in significant amounts.

As we have seen, both PCBs and dioxin are mixtures of closely related chemical molecules. In recent years, determining the toxicity of the different molecules composing these mixtures as well as the toxicity of commonly encountered combinations of these molecules has commanded high governmental priority. Still the risk estimates of the different possible mixtures remain uncertain. The focus on these two chemicals has given impetus to the development of broad data bases concerning many other chemicals that could help in clarifying risks from various mixtures that have differing ratios of the same ingredients.

In addressing mixtures, policy officials usually assume the harmful effects of one ingredient must be added to the harmful effects of other ingredients to understand the overall hazard of the mixture although the methods of addition are not clear. For example, the presence of 10 ppm of one toxic chemical and 10 ppm of another will induce a harmful effect which is more severe than the adverse effects induced by either of the chemicals acting independently, according to expert advisers.

The concept of addition seems reasonable in the absence of better information. However, when the purpose of risk assessment is to support regulatory actions in numerical terms, crude estimates based on simple addition are hardly adequate. If the economic stakes are sufficiently high that proposed regulations are challenged by the affected parties, the advocates of simple addition will be hard pressed to defend their risk estimates. The methodology for estimating risks to humans and to ecological resources from exposures to mixtures is clearly a neglected area of the environmental sciences.

Chemical Threats to Ecological Resources

Another poorly understood, but critical, aspect of risk assessment is the hazard to ecological resources posed by some man-made chemicals. In the laboratory, some of the effects of one or a few chemicals on several types of plants, fish, or vertebrates can be studied under highly controlled conditions. Numerous studies of the ecological toxicity of individual chemicals have been carried out in artificial streams, aquariums, and greenhouses for many years. But in nature, many chemicals of both natural and human origin are constantly disrupting a wide variety of flora and fauna, and more extensive field studies are essential to extend the laboratory investigations into real-life situations.[10]

Some of the more obvious ecological effects of man-made chemicals can be easily observed, such as the dieback of forests attributable to air pollution and the disappearance of fish in polluted streams located near industrial complexes. However, many subtle but significant chemical interactions within highly integrated ecological systems remain cloaked in mystery, particularly in areas where pollution has not simply overwhelmed every form of biological life. For example, the effects of pollutants on the reproduction of species, on the competition among aquatic organisms for nutrients, and on the susceptibility of plants to predators and disease cannot be adequately tested in the laboratory and cannot be readily observed in the field.

In some areas of the country the ecological impacts of toxic chemicals are very important for the future survival of individual species or of entire ecosystems. In other situations, their significance is lessened by much worse woes: bulldozers leveling the land, careless campers igniting forest fires, and developers draining marshlands.

The changing character of Lake Mead, the recreational area behind Boulder Dam on the Colorado River, illustrates the difficulty in determining whether ecological changes are for better or for worse. While the changes were not induced by toxic chemicals, this case study vividly illustrates the trade-offs involved in tampering with nature.

In 1987 and again in 1988 local scientists organized a fleet of more than 300 small motorboats to fertilize the lake. Each year they dumped more than 130 tons of ammonium polyphosphate fertilizer into one of the arms of the lake. The scientists wanted the fertilizer to serve as a nutrient for the lake and to increase the amount of algae in the lake. They claimed success. Why did they want algae which had been the scourge of Lake Erie 15 years earlier and had clogged many other inland aquatic reserves?

In the early 1980s, the city of Las Vegas had installed a tertiary sewage treatment system at the insistence of the EPA to reduce the flow of nutrients into the lake. Previously these flows of wastewater had carried biologically active materials which contributed to the nutrient loading of Lake Mead. Meanwhile, nutrient levels in the lake had already been declining due to the trapping of naturally occurring sediments behind the new Glenn Canyon Dam upstream on the Colorado River. Thus, two sources of nutrients for the lake were being cut off. But nutrients lead to algae, and reducing the levels of algae in the nation's lakes had become a major environmental objective. Weren't these approaches contributing to this objective?

However, with the disappearance of the algae, the lake had become too clean. While water-skiers enjoyed sparkling aquatic clarity, the more important sport fishing industry was in serious trouble. The striped bass and rainbow trout which depended on an abundance of nutrients were fast disappearing. Thus, out went the cry for the scientists to help restore the biological producuvity of the lake. Today the fish seem to be returning, and the water-skiers are still out in force every weekend. Nevadans are learning how clean is "clean enough."[11]

One of the most difficult aspects of ecological risk assessment is the identification of "end points"—those types of observable effects from pollution which indicate the onset of irreversible damage. Such damage may be occurring within individual cells; it may be affecting the composition of an entire community of a species; or it may be disrupting an ecosystem. More likely, destruction is occurring at several levels at the same time. With regard to animal species, some scientists have suggested that the destruction of habitats where reproduction takes place, such as nesting areas, may be the key to identifying serious ecological damage. As to broader ecosystem concerns, scientists have proposed that changes in the quantities and types of organic matter which are naturally regenerated, for example in a forest, are a salient indicator of serious ecological impacts which may be attributable to pollution.

Ecology has been called the science of resiliency. If given a chance following the destructive practices of man, streams will recover, forests will regenerate, and wildlife will come back. However, at some point, pollution stretches nature beyond its limit. Like a rubber band, biological resiliency has a breaking point. When does chemical pollution push biological systems to that breaking point? This is a major challenge for ecological risk assessment.

As to the "value" of nature, a few years ago the Congress began to require assessments of damage to natural resources that resulted from hazardous wastes. Such assessments can play an important role in financial settlements at Superfund sites. Determining reasonable financial losses due to ecological damage is not easy. The lost incomes associated with fish that are destroyed, trees that are damaged, and recreational visitors who stay away from polluted areas are very important to a local economy in the short term. However, a more comprehensive measure of loss that has been used in court actions is the decline in the sale value of properties in the affected areas—current values in comparison to values of similar property in nearby unaffected areas. This approach, while far from perfect, attempts to integrate many indirect costs associated with ecological damage.

During the past two decades our nation has not given adequate attention to ecological risk. The experts often don't even have a good starting point. They don't know the state of ecological resources, and trends in the conditions of our forests, our vegetation, and our animal populations have not been adequately established. Too frequently, scientists must rely on anecdotal evidence and on a never-ending series of complaints by interested parties as resources continue to be destroyed.

Recently the EPA launched a major nationwide monitoring program to begin to improve appreciation of the condition of the nation's land and water on a scientific basis. The program is targeted specifically on near-coastal waters, forests, freshwater wetlands, surface waters, agroecosystems, and arid land. In each type of environment a number of questions will be addressed. What are the key resources that deserve immediate attention, and what type of damage have these resources already endured? Are the extent, magnitude, and location of the damage changing? Is the damage related to pollution or to other disturbances? What level of uncertainty is associated with the estimates? What can be done to reverse undesirable trends, and what are the costs? A monitoring program will not fully answer any of these questions but will certainly help improve the basis for managing natural resources that cannot be easily replaced.[12]

Since the mid-1970s, the EPA's priorities have been tilted toward reducing threats to public health from man-made chemicals while paying far less attention to ecological concerns. However, health and ecological aspects are intertwined, particularly for future generations of Americans who will increasingly disperse into the most isolated corners of the nation. Techniques for assessing ecological damage of the past and predicting future impacts of uncontrolled chemical activities will be desperately needed in the years and decades ahead. No longer should they be considered only of secondary importance because of the urgency of public health problems.

Documenting Uncertainty Leads to Informed Decisions

The previous sections have considered human health risks from exposure to carcinogens, risks posed by mixtures of chemicals, and risks to ecosystems from chemical releases into the environment. While scientists use very different approaches to estimate these types of risks, almost all risk assessments are surrounded by high levels of uncertainty.[13]

One senior government scientist has noted that the difference between risk assessment and five-year weather forecasting is that at least with the weather forecast, if you wait five years, you find out if you were right.[14] Or in the words of former EPA Administrator William Ruckleshaus, risk analysis ". . . is a kind of pretense; to avoid paralysis of protective action that would result from waiting for 'definitive' data, we assume that we have greater knowledge than scientists actually possess and make decisions based on these assumptions."[15] At the same time, he was a vigorous supporter of detailed risk assessments based on available information, no matter how sparse, and of efforts to improve the methodologies for carrying out such assessments.

Regulators and litigators need numerical limits that reflect risk judgments—numbers which can be used as the basis for setting tolerable limits of exposure, for determining when polluters are posing a threat to society, and for demonstrating before the courts that cease and desist orders are needed. In response to this demand for clear estimates of risk, scientists continually stretch the limits of their knowledge, interpolating and extrapolating at every turn. They have no alternative.

Scientists must be prepared to defend their scientific judgments not only before their peers but also before government officials, hostile opponents, the media, and the public. Why was a safety factor of 1000 and not 100 used in extrapolating the effects of chemicals on mice to those on humans? Why were laboratory experiments showing a chemical affecting mice at very high dose levels considered more important than studies of workers exposed to the same chemical who had no adverse reactions? Why are outdoor air pollution levels given so much weight when most people spend 22 hours each day indoors? How can scientists talk about the impact of acid rain on 200,000 American lakes when they have made measurements in only 2000 and every lake is unique?

The list of questions concerning the soundness of risk assessments seems endless. Thus, many judgments must be made en route to a recommended numerical limit. These judgments should be clearly documented and explained in language understandable to both scientists and nonscientists. In some cases, debates will ensue as to whether judgments are scientific or are laden with policy viewpoints. In any event, placing the full rationale for risk estimates in public view should lead to decisions that best serve societal interests.

The Times Beach incident of dioxin contamination mentioned earlier illustrates that judgments can be important in developing a risk assessment. Many uncertainties persisted concerning the health effects of dioxin on laboratory animals, let alone humans, and only guesses were available as to the frequency and intensity that children would play in contaminated soil and absorb dioxin through the skin or by licking their fingers. Different scientists could easily have concluded that an appropriate action level should have been 100 parts per trillion (ppt) or 100 parts per billion (ppb) rather than 1.0 ppb. Then a responsible policy official should decide how numerical limits are to be used in the best interests of society—interests that transcend science.

Yet the recommended action level was presented by the toxicologists as a single number. Little effort was made to document the uncertainties surrounding the number in a manner that permitted the choice of any level other than 1.0 ppb. The scientists decided themselves that the correct number was 1.0 ppb. They did not provide a framework which would have allowed the officials responsible for setting the limit an opportunity to examine the implications of alternative numerical limits, in terms of health effects and economic and social consequences.

In short, analysts can seldom hope realistically to determine the precise risk associated with a chemical or a mixture of chemicals. Rather, the analyst can articulate the extent of knowledge about the hazard of the chemical and the uncertainties associated with this knowledge. This information should be presented in terms of a range of numerical limits within which the true hazard is likely to fall. If the scientific information about the chemical is quite complete and of high quality, the range will be narrow. If the data are sparse, the range will be broad.

Science, Values, and Environmental Priorities

During the past few years, the EPA and other environmental regulatory bodies at the federal and state levels have become sensitive to the fuzzy dividing line between science and policy as was illustrated in the case of dioxin. They have tried to separate the process of assessing risks from the decisions of whether and how to reduce risks. Proponents of such separation argue that risk assessments should be based only on science whereas regulatory decisions, while drawing on the assessments, must employ a different type of analysis—a balancing of political, economic, and social factors within legal constraints.[16]

This second type of analysis must ask, for example: How much can society afford to spend to curtail this risk? How quickly should and can the risk be reduced? Should all polluters be required to take the same actions, or should some, such as small business, be exempted? What precedents will be set by this regulatory action, and what are the future implications?

The vigorous advocacy of separating science from social policy to make regulatory decisions has helped clarify the boundaries between scientific facts, scientific judgments, and value judgments. However, this concept of separation should not be pushed too far since the dividing lines are seldom completely clear. Value judgments which transcend science are inevitably encompassed in assessing risk, with the choice of safety factors being a case in point. Furthermore, articulating the uncertainties surrounding risk assessments in a manner which can be understood by policy officials is seldom easy. Invariably, decision officials want to know how uncertain is "uncertain." In response, scientists frequently develop statistical measures of uncertainty, only to be told that they should be more precise in the future.

Sometimes, important inhibitions to separating science and values arise as a risk assessment moves into the political arena. Many policy officials are most comfortable when making decisions based on "objective" analyses which can be set forth on computer graphs and printouts. They are not eager to open up debates on the assumptions which could undermine the objectivity of "scientific" risk assessments. Very simply, they may look at a risk assessment as a crutch to justify a political decision rather than as an aid in reaching the decision.

Clearly, separating facts from judgments should be encouraged to the fullest extent possible. Since complete separations are seldom possible, policy officials should participate directly in scientific assessments as necessary to understand the process and the assumptions. Also, scientists should play a continuing role as these officials reach their regulatory decisions which often rest, or are said to rest, on an appreciation of scientific assessments and uncertainty.

Regardless of the imperfections in the risk assessment process, I have repeatedly stressed that quantitative assessments have enormous value. In addition to providing a springboard for reaching regulatory decisions, they help government agencies and other institutions set their priorities. Funds are never available to address every risk situation immediately. If one analysis indicates that perhaps one dozen people may be at risk from an environmental problem and a second assessment of another situation suggests that as many as 1000 lives are in jeopardy, officials have a basis for deciding how to divide their resources in addressing the two situations.

At the same time, we must be wary of those who argue that until "major" risk problems are resolved, the government shouldn't divert its attention to correcting related "minor" problems. For example, they may ask, "If there are multiple polluters along a stream, is forcing one of the minor polluters to limit discharges before the major polluters take corrective action fair?" Or, "Should the public be concerned with chemicals entering into surface waters from the discharge pipes of industry when in the same region 80% of the contamination of the same water bodies is coming from agricultural runoff?"

Obviously, the largest polluters should be the major targets for enforcement offices. But even small amounts of some very toxic chemicals can be harmful. The regulatory process has become so complicated that the only sensible approach is to address simultaneously as many environmental hazards as possible with priorities tilted toward the greatest risks as necessary. Waiting to address all risks seriatim on the basis of "worst polluter first" may delay important corrective actions for decades. Indeed, resolving minor problems often places greater pressure on the major polluters to clean up their acts.

As early as 1970 President Nixon presented a then popular and seemingly rational approach for addressing environmental risks in his statement accompanying the reorganization of the federal environmental effort:

> . . . for pollution control purposes the environment must be perceived as a single interrelated system. A single source may pollute the air with smoke and chemicals, the land with solid wastes, and a river or lake with chemical and other wastes. Control of air pollution may

> produce more solid wastes which then would pollute the land or water. . . . A far more effective approach to pollution control would identify pollutants; trace them through the entire ecological chain, observing and recording changes in form as they occur; determine the total exposure of man and his environment; examine interactions among forms of pollution; [and] identify where on the ecological chain interdiction would be most appropriate.[15]

Such sophistication was too difficult to translate into practical terms and quickly gave way to more pragmatic steps of simply turning off the obvious pollution sources which were contaminating streams and clouding the air. This indeed was the approach the government adopted for a number of years following the 1970 statement. However, many toxic chemicals which are far less obvious than chimneys belching smoke eluded almost everyone. Indeed, toxic problems remain largely invisible, and their impacts are often delayed for decades. Thus, returning to a more comprehensive analytical approach to help guide the way in addressing the risks posed by chemicals seems appropriate. Still, care is needed to ensure that shortcomings in mathematical models and scientific data bases do not become excuses for inaction until the totality of the problem is better defined.

During the 1990s the United States will spend tens of billions of dollars each year to prevent and to clean up chemical problems in the environment. While procrastination in controlling toxic chemicals cannot be tolerated, the nation must use its financial resources wisely. Comparable attention must also be directed to other problems that erode the natural environment such as unrestrained development of estuaries and wetlands which often can be far more destructive than chemical pollution. In short, we need well-structured decision frameworks that can help in balancing the multiplicity of environmental, economic, and fairness concerns attendant to environmental regulatory decisions—concerns that reflect societal interests in the broadest sense of the term.

The Judiciary Speaks Out on Risk and Uncertainty

Since the birth of the EPA, the decisions of the Agency to regulate environmental chemicals have been under close scrutiny by the judiciary throughout the country. Environmental groups repeatedly petition the courts to force the EPA to strengthen regulatory approaches which they consider too timid. Individuals harmed by toxic chemicals turn to the courts for compensation. Conversely, regulated parties frequently seek redress for proposed regulations which they believe are based on exaggerated estimates of risk and which they contend will have disastrous consequences for them.

The U.S. Court of Appeals for the District of Columbia Circuit, in particular, has been seized with reviewing and ruling on the appropriateness of risk-based decisions of the EPA. In a very perceptive and well-reasoned speech in 1980, former Senior Circuit Judge David L. Bazelon of that court provided incisive commentary on the problems facing both the government and society in addressing environmental risks. He noted, for example:

> . . . the electorate must have an opportunity for the final say about which risks it will assume and which benefits it will seek. Elitists will say that most people are incapable of evaluating risks. Such a claim has no more place in an agency's decision-making than in an individual's choice about health care. Experts who are beyond reach and beyond view must never be allowed to arrogate those decisions to themselves.
>
> But what if an agency lacks the knowledge to state risks with certainty? For some activities, the magnitude of potential harm and the probability of its occurrence may be essentially unknown. . . . Risk estimates may depend on future contingencies of human behavior or other highly complex and unpredictable variables. . . . The best risk estimates are subject to an unknown degree of residual uncertainty and may thus overstate or understate the dangers involved. Many times, however, an agency must act in circumstances that make a crap game look as certain as death and taxes.
>
> . . . Perhaps those who seek to conquer uncertainty do not see eye to eye with those who act in spite of it. A "pure" scientist is usually acutely aware of the tenuousness of his assumptions, the competing interpretations of the data, and the limits of his knowledge. He presses outward upon the line between the known and the unknown. He does not resist disclosure: indeed, his career advances through it. If anything, the scientist is more likely to overemphasize uncertainty than to hide it.
>
> Those who must make practical decisions, on the other hand—regulators, physicians, engineers—cannot always afford science's

> luxury of withholding judgment. Indeed, they may be tempted to disregard or even suppress uncertainty. Uncertainty is messy. It cannot be stated as an objective quantity or factored into a decision as if it were a risk of known probability. Decision makers must consider data from many disciplines. Uncertainty detracts from simplicity of presentation, ease of understanding, and uniformity of application.
>
> To focus on uncertainties is to court paralysis: to disclose them is to risk public misunderstanding, loss of confidence, and opposition. Even though some uncertainty is inevitable, pointing it out will always create pressures for "just one more study." And yet, the decision maker knows too well that delay is also choice, with risks of its own.
>
> I am told that instead of disclosing uncertainty, decision makers may want to compensate for it by intentionally inflating risk factors. Engineers and physicians likewise choose to build in safety margins and err on the side of caution. I do not criticize those "conservative" decision rules; indeed, where health and safety are concerned, they are the only ones that make sense. But such rules cannot erase the uncertainty inherent in many decisions.[17]

Judge Bazelon has left behind a legacy of sophistication in judiciary understanding of the day-to-day problems faced by regulatory agencies in attempting to balance the many factors that comprise societal interests.

Of course the courts have also become a battleground for people who have been injured by environmental chemicals and seek financial compensation. They enter claims against the government, alleging that federal or state authorities have not done their jobs in providing protection from exposure to environmental chemicals. They also sue private companies that are responsible for the manufacture, distribution, or disposal of chemicals. Causation—relating the injury to the action of the defendant—is almost always a central issue in these court hearings. A second issue is the amount of compensation that is appropriate once causation has been established.[18]

During the past decade several courts have devoted considerable attention to allegations of government negligence in handling nuclear materials. While the analogy between nuclear radiation and toxic chemical problems is far from exact, these radiation cases have been instructive in clarifying the concerns of the courts when considering risk issues and in determining compensation levels for victims of hazardous exposures. In one case in the mid-1980s, the judge priced the life of a person at about $500,000. Using this figure as a baseline, he ordered the government to pay the survivors of individual cancer victims from radiation exposure sums in the hundreds of thousands of dollars depending on the age, earning capabilities, and general value to the community of each victim.[19]

The risks being weighed by the courts in compensation cases relate to past exposures of specific people to specific chemicals. In both radiation and nonradiation cases, a particularly difficult aspect for the claimants is reconstructing and documenting their past personal behavior patterns over periods of up to twenty years. These patterns should persuasively show the degree of contact the claimants had with the chemical of concern. Sometimes, such as in the case of residents of industrial areas subjected to pollutants from many sources, behavioral patterns also reveal exposures to other injurious chemicals as well. In addition, the problems of relating a specific type of exposure to the particular injury—for example, lung cancer, leukemia, heart disease, kidney dysfunction—are formidable since such diseases can result from many causes.

Reflecting frustrations in this area of causation, one popular legal doctrine now holds that each contributing party should be fully liable for injuries that result from a chemical problem. For example, three companies may manufacture a product which is shown to have injured a particular consumer. In this case it is impossible to distinguish which company's product caused harm to the consumer due to his random selection of the brands over a period of many years. Thus, the court would rule that each of the companies can be held liable for the injury. In another type of case, if several companies have disposed of chemical wastes in a particular dump and the wastes become comingled, each company can be held liable for any of the problems caused by the dump.

Two recent dramatic cases show the magnitude of the claims recently filed in the courts by people who had been injured from chemical exposure problems.

In Bhopal, India, often called the chemical industry's Three Mile Island, a toxic gas used in the production of pesticides escaped from a manufacturing plant owned by the Union Carbide Company and blanketed a densely populated urban area. The death toll exceeded 2000, and thousands more were blinded or maimed for life. The source and the effect of the release of the gas were never in dispute. Union Carbide abandoned its early efforts to shift the blame to its Indian affiliate and decided to settle the claims, initially totaling several billion dollars, for $400 million, a small price to pay for the human suffering that resulted.

The claims of shipyard workers and others who have handled asbestos for many years against the Johns Manville company and several other manufacturers of asbestos are reaching billions of dollars. Eventually many of these claims will probably be settled. However, reflecting the difficulty in proving that asbestos was the cause of cancer and related lung injuries, a large portion of the payments will go for legal fees with one estimate being that the victims and their survivors will receive less than one-third of the total payments.

For many years the Congress has debated the desirability of dictating wide-ranging compensation for victims of exposures to toxic chemicals. The long-standing law providing for government arbitration in resolving compensation cases of black lung among coal miners is often cited as a possible model. However, applying the precepts of this legislation and other approaches to worker's compensation to the problems of compensation for exposure of the general population to environmental chemicals needs to be approached with great caution. There are relatively few coal miners, but there are tens of millions of Americans who are exposed to trace levels of toxic chemicals.

Clearly, the courts are dealing with issues which should be of concern to all officials responsible for the control of toxic chemicals. However, many of their decisions are retrieved from law libraries only as new cases reach the courtrooms. Their rulings deserve much greater immediate attention from everyone involved in preventing as well as reacting to toxic chemical problems. Dealing with claims which are tied very directly to personal harm is an excellent way to underscore the many aspects of chemical risks, and particularly the many uncertainties.

Reducing Risks during the 1990s

In 1850, the average life expectancy in the United States was less than 45 years. Now it is over 70 years. The industrial revolution was at the core of this change in the health of the population.

Living in the industrial age we have become preoccupied with types of risks that didn't seem important a century ago, including risks posed by some man-made chemicals. All institutions—regulatory agencies, industrial companies, research laboratories—have a responsibility to join in the national effort to reduce chemical risks. Even if they all strive toward this goal, some level of risk will persist. The definition of "safe" in *Webster's Ninth New Collegiate Dictionary* as "without risk" is obsolete. During the next decade, an important challenge facing our institutions is to develop mechanisms which will permit all elements of society to participate more fully in deciding the levels of safety, or "acceptable" risk, for specific situations.

Given the embryonic state of risk assessment, a key in responding to this challenge will be major educational efforts—efforts to improve the expertise of specialists and the sophistication of the public. All participants in the national environmental effort need to understand better the potential and limitations of science and how to incorporate high levels of scientific uncertainty into responsible public policy. Education about risk must begin in the schools, receive far greater attention in universities and colleges, and continue into the town meetings where practical problems are addressed.

As this educational process proceeds, public perceptions of risk will continue to be a driving force in environmental protection programs. Risks that are difficult to understand, and particularly the probability of harm from intermittent exposures to low levels of chemicals, are often much more threatening to the public than more familiar, and even more serious, risks such as automobile accidents. Scientific studies will help in reducing the mystique of chemical risks. Also, improved estimates of the economic costs in reducing risk levels to zero will convince some people that certain risks may not be as bad as they seem. Still, the public wants to be exposed to a minimum of involuntary risks, and chemicals in the environment will remain an easy target for public outrage.

Meanwhile, concerned citizens grudgingly accept the argument that every chemical problem cannot be solved immediately. They want to be certain, however, that the procedures for determining priorities and for resolving each problem will be fair. The public needs to be constantly reassured that governmental agencies are responsible and free from scandals and favoritism. Unfortunately, a few regions of the country, such as those embroiled in debates over disposal of nuclear waste, may never see the day when the public believes there is veracity in governmental pronouncements about environmental risks.

Increasingly, risk assessments, whether sponsored by government agencies or by local institutions, must stand up as "objective" under rigorous scrutiny by all concerned parties. The assessments can then play important roles in educating policy makers, in determining priorities, and in reaching conclusions on the most appropriate actions to reduce risks. However, scientific studies which blur the known facts with untested hypotheses, which fail to clearly articulate the uncertainties, or which subsume and hide value judgments are best left in the desk drawers of their authors.

At the present time, hundreds of institutions and thousands of specialists are busy assessing chemical risks. These experts work for federal and state agencies and their contractors. They are employed by insurance and financial institutions. They are full-time researchers at universities and think tanks. They are specialists engaged by labor unions and professional associations.

Some of these risk specialists are attempting to improve the methodological approaches to risk assessment. Frequently they share their experiences at scientific meetings and through the pages of scientific journals. However, much experience is simply forgotten after each problem is solved. Newcomers streaming to the field usually spend many months and years climbing the learning curve, often unaware of pitfalls encountered many times before by their predecessors.

In financial terms, government and industry spend billions of dollars each year developing scientific data which can be used in assessing chemical risks. The federal government alone earmarks tens of millions of dollars annually for improving risk assessment techniques. Given these large expenditures and the currently fragmented efforts for assessing risks, a modest investment by the Congress of a few million dollars each year to provide an effective focal point in Washington to coordinate and improve the risk assessment process on a nationwide scale seems long overdue.

The EPA and other agencies will argue that they are already the focal points. But each speaks with a different voice. None adequately recognizes the vital roles that state agencies, private sector organizations, and the courts should play in setting the national direction for assessing risks. A new congressionally mandated forum, with representation from a broad spectrum of decision officials and other interested parties within and outside Washington, could play a useful role in reaching a consensus on how to approach risk decisions. They could sponsor educational and information-sharing activities that in a relatively short time would have considerable payoff in reducing risks and saving dollars.

We will see in a later chapter that the most effective way to reduce future risks from pollutants is to stop the generation of pollutants in the first place: for example, through the use of more efficient manufacturing processes, by the promotion of energy conservation measures, and through reduced reliance on chemicals known to cause problems, such as pesticides. A second line of defense is recycling and reuse of products and wastes which cause trouble—including the remelting of metals and the reclaiming of solvents, for example. If wastes still persist, then more effective controls on liquid effluents and atmospheric emissions and tighter containment of solid refuse are clearly needed.

While industrialists and agriculturalists can reduce the generation and spread of chemical pollutants, they will not be able to reduce leakages into the environment to zero. Also, we must cope with the chemicals which are already in the countryside. Therefore, identifying and assessing chemical risks will continue to undergird environmental protection efforts for the indefinite future.

Finally, the uncertainties surrounding risks from chemicals are enormous. At the same time, the costs of placing restrictions on commercial activities which are responsible for such uncertain risks can be substantial. But the cost of not taking action to reduce risks can be devastating in the long run. The experts often debate the cost of scrubbers and incinerators, of cleaning up spills and waste sites, and of developing nontoxic substitute products. They discuss far less frequently the increased cost of health care for pollution victims. They neglect the cost of finding alternative drinking water sources after aquifers are polluted. They forget the lost productivity and the beauty of land abandoned to contamination.

4 Explaining Risks to an Aroused Public

I know of no safe depository of the ultimate powers of society but the people themselves; and if we think them not enlightened enough to exercise their control with a wholesome discretion, the remedy is not to take it from them, but to inform their discretion.
—Thomas Jefferson

Advocates of bad policies sometimes imagine that they can get away with anything if they sell it cleverly enough, while advocates of good policies sometimes imagine that they don't have to sell at all.
—U.S. Environmental Protection Agency

Even though good risk communication cannot always be expected to improve a situation, poor risk communication will nearly always make it worse.
—National Academy of Sciences

The Power of Television

"Would you live near a plastics plant manufacturing polyvinyl chloride?"

For 90 minutes an 11-member CBS team of reporters and television cameramen surrounded me in my EPA office as the interviewers over and over again tried to pin me down with different phrasing of the same question. For 90 minutes I avoided a direct answer. The EPA Administrator commended me for my oratorical skill. But my friends condemned me for the bureaucratic double-talk that came through during the 90-second clip from the interview shown that evening.

At the time of the broadcast in 1974, I surely would not have lived near a polyvinyl chloride plant which might expose my two young daughters to vinyl chloride that had been detected seeping into nearby neighborhoods. Since I had the option to live elsewhere, I simply would not have taken a chance regardless of how low the risk seemed to the experts. At the same time, I, as a government spokesman, could not express this view on a nationwide news broadcast. Such a statement could trigger a flurry of panic among the millions of Americans who had lived near these plants for many years and who did not have the financial wherewithal to move from their neighborhoods.

During this interview and my subsequent meetings with the press, the reporters were not interested in estimates of risk probabilities associated with vinyl chloride exposures. They didn't want to hear about the difficulty in extrapolating from laboratory experiments to real-life situations. They wanted a simple answer to a simple question. Was an EPA official who was at the center of the government's investigation of the problem of vinyl chloride releases prepared to trade places with the residents of Louisville, Long Beach, or Painesville who lived near these plants? If not, what was the government going to do to protect these residents?

The power of television is at its height when ordinary people painfully convey their individual stories and when personal tragedies are laid at the doorstep of the government which has a responsibility to protect all citizens from harm. Since environmental risks affect people from all walks of life, the possibilities for compelling stories are endless. Thus, the impacts of chemicals on people—and also on fish, on wildlife, and on forests—will continue to provide good grist for television, both nationally and locally, during the years ahead.

An authoritative study of national television coverage of environmental risks reached some interesting conclusions:

> ABC, CBS, and NBC's carefully crafted and expensively produced evening news broadcasts devoted 1.7 percent of their air time to 564 stories about man-made environmental risks during the period from January 1984 to February 1986. Little relationship was found between the amount of coverage and public health risk. Indeed, the networks appeared to be using traditional journalistic determinants of news (timeliness, proximity, prominence, consequence, and human interest) plus the broadcast criterion of visual impact to determine the

> degree of coverage of risk issues. . . . Given the media's need for news pegs, acute and chronic risk stories were covered differently. Acute risk stories were reported in a clearly defined cycle, peaking on the second day with on-the-scene reports and film clips of devastation. In keeping with a decrease in visual drama, later reports were shorter and emphasized legal and political considerations. Chronic risk coverage followed the release of new scientific, legal, or political information.[1]

Another study during the 1980s of coverage by the major television networks and three leading national newspapers of several major environmental problems involving toxic chemicals concluded that "media coverage of chemical health risks is likely to reflect the assumption that a risk is serious enough to require action; uses scientific data sparingly; and presents a sensationalized perspective."[2]

As to the experts who are shown on television, both studies concluded that government spokesmen are usually the primary sources of information with scientists and independent experts consulted only sparingly.

Many government and industry officials—as well as environmental groups—firmly believe that the public does not know when to worry and when to relax, and they like to blame television for this confused state of mind. These same critics must also assume some of the blame themselves. They are frequently less than open, forthright, and honest in their dealings on television. Indeed, my debut on CBS in grappling with vinyl chloride was not a proud moment.

Most government and industry officials feel confident with well-rehearsed appearances arranged to sell a point of view, but they are far less comfortable in unstructured give-and-take sessions designed to bring all the facts and opinions out into the open. They know that vocal elements of the public harbor a deeply embedded mistrust of government policies and industry motivations. Nevertheless, almost all senior environmental officials from government and industry face the cameras at one time or another. Some recognize their clear obligation to cooperate with the media which has become a major environmental force nationally and locally. Others realize that they have no place to hide and have become accustomed to going through the motions of communicating with a skeptical public. A few public officials genuinely welcome public debates of issues they are addressing and carefully weigh questions and comments of their critics in carrying out the assessments leading to their decisions.

The public often has difficulty distinguishing sincerity from cynicism on the part of officials. Television time is carefully husbanded, and more questions are often raised than answered in the brief public sparring. Thus, short clips frequently reinforce stereotypes of the "typical" government official in the eyes of those who have strongly held views on issues that are raised on television.

In response to the general awareness of the power of television, a call for the training of risk communicators is now sweeping through Washington. Ten percent of all EPA employees are being trained in this field—an unprecedented level of interest in communications among government agencies or indeed among any workforce. High-priced public relations firms are in great demand by industry to organize classes for the community outreach specialists of the large companies. Material on how environmental officials can effectively communicate with the public is flooding the popular and scientific literature. Hopefully, these officials will pay heed to the admonition in all the literature that the most important aspects of communication are honesty and forthrightness.

In 1987, I was swept up in the initial wave of risk communication training. I was chosen as one of five members of the American Chemical Society to receive practical instruction on how to provide the public with more objective information about chemical risks. All five of us felt that chemicals were receiving a bad rap, particularly in Washington, in being depicted only as an enemy of society when in fact they are critical to our very survival. At the same time, none of us worked for the chemical industry.

We spent one day in a Washington television studio with a seasoned network television commentator. She confronted us with all the popular clichés about killer chemicals and poisoning of the environment. She demanded immediate answers to her questions and would not accept scientific gobbledygook. She was deliberately abrasive and nasty, but she was effective. We had great difficulty surviving her well-directed barbs.

After this humiliating experience, I assembled many one-liners that can be injected amid rapid-fire questioning to help put chemical risks into perspective: "When properly handled, chemicals are safe." " 'Risk' is the extent of the threat to human health and the environment when chemicals are not properly handled." "Many naturally occurring chemicals present greater risks than man-made chemicals." "The risks of cancer and other effects from chemicals in the environment are much less than the risks from automobile accidents, smoking, and drugs."

However, I don't think I will ever adjust to her contention, and the contention of many other experts, that a specialist's effectiveness on television depends largely on speaking mannerisms (50%) and body language (40%), and only secondarily on the substance of the remarks (10%). In any event, television is clearly a dominant force in communications between the government and the public on environmental issues. Increasingly, mastery of television skills is becoming a decisive determinant in the shaping of environmental policies.

The Continuing Impact of the Press

Turning to the printed word, almost every environmental official has bridled at inaccurate, incomplete, or distorted reporting of his or her statements. Many have been disappointed with the press for ignoring the "authoritativeness" and the importance of their words and not giving greater coverage to statements which support their views. While the impact of television close-ups of chemical fires, of chemical leaks and spills, and of suffering victims can shock the public, the less visual newspaper accounts of environmental problems also can stir emotions of the man on the street and rile the anger of responsible officials.

Veteran environmental officials often personally enjoy being quoted in the press. Almost every environmental agency has a staff which screens the press and clips articles reporting on the activities of its leaders. Too often, however, environmental officials become very cynical about press reporting. They try to conceal their belief that environmental reporters are novices in need of an education, and occasionally they simply talk down to the press.

Practicing scientists and other specialists dealing with the details of environmental problems sometimes view the press as their only effective vehicle for communicating with the politicians who make decisions. Also, environmental lobby groups heavily orient their activities toward media appeal, and they frequently measure their success in terms of inches of press coverage nationally and locally.

Of course many newspapers and magazines publish excellent reports of the histories, risks, and uncertainties of chemical incidents. A number of environmental reporters are highly skilled in identifying significant scientific findings and uncovering crucial environmental assessments that for one reason or another were never released to the public.

While reports on environmental crises tend to dominate the coverage, important policy developments and general assessments of the state of the environment are not neglected. For example, on September 8, 1989, while in Davis, California, I bought copies of the *New York Times, Los Angeles Times,* and *Sacramento Bee* for a group of visiting dignitaries. On the front page of each of these papers was a lengthy account of the findings of the National academy of Sciences in Washington that American farmers and farming enterprises used excessive amounts of fertilizers and pesticides. An important study of the academy had concluded that federal agricultural subsidy programs designed to encourage more efficient and profitable agricultural approaches had also encouraged excessive use of chemicals. A number of case studies of alternative approaches to farming showed that farming which is less dependent on chemicals can often be more successful in raising productivity per acre.[3] The press accounts reporting these findings certainly had an impact in the farming communities of California as well as in Washington, and I suspect in many other communities throughout the country.

While many articles and popularized books present balanced views of environmental issues, catchy headlines are still used to aggressively sell publications. Thus, the *Miami Herald* banners on the front page, "New migraine for motorists: mandatory testing of exhausts." On an inside page, the headline reads, "Tailpipe test will be pain for drivers." Only reluctantly does a subheadline state, "But cleaner air could be the result."[4] Similarly, Americans are regularly exposed to dramatic titles designed to promote the sale of books. Recent titles include *The Ozone Crisis, Toxic Terror, Laying Waste: The Poisoning of America by Toxic Chemicals, The Poison Conspiracy, Malignant Neglect,* and *The End of Nature.*

Sensationalized headlines and titles will remain an important dimension of the environmental movement. They will continue to catch the attention and often distort the views of the public. Hopefully, the reporting that follows the headlines and the discussions behind the titles of the books will be more responsible. We simply will have to live with the reality that dramatic phrases sell newspapers and books. Only an expanded educational effort on a broad front can create a more sophisticated readership which recognizes that complicated environmental issues cannot be presented in a handful of words.

Seeking Consensus but Encountering Controversy at Public Meetings

While the media will undoubtedly remain the major communications link between government agencies and the general public in the environmental field, other important forms of communication are rapidly spreading as some segments of the public become deeply immersed in controversies. Often the problems are so complex that even the most articulate journalist has great difficulty in presenting the key issues within the constraints of press deadlines and limited space. Also, much of the supporting information is highly technical and hardly suited for television or newspapers.

Nonetheless, government officials should not underestimate the capabilities of environmental reporters or the sophistication of the public. When an environmental problem has a direct impact on a community or on individuals within the community, local newscasters and residents quickly become quite expert in many detailed aspects of issues previously reserved for only specialists. Even in those cases when a government agency decides to shoulder the entire burden of telling its own story through seminars with local community leaders, through public meetings, and through volumes of technical documents, it should recognize the media as an important interpreter of developments in communicating with the public.

In recent years, meetings between government officials and the public on environmental issues have become commonplace at the federal, state, and local levels. Laws or regulations may require such public sessions. Interest groups may demand them. Meetings may be organized by government officials who are seeking support or who are trying to minimize opposition concerning their proposals. Some officials may initiate interactions with the public because of their personal commitments to the concept of a public role in decision making.

Public meetings can be highly structured and carefully orchestrated events. They can be informal and freewheeling. They can be held in connection with seminars involving experts. They can be organized as focus groups involving parties who have similar or conflicting points of view.

Environmental controversies are very diverse. The political setting for each problem is usually quite unique, and the public's reactions are highly dependent on deeply engrained local social and economic interests. Several unusual personal experiences have influenced my views as to the role and impact of public meetings. These case studies highlight a few important aspects of seeking consensus or at least common understanding. They may be helpful in identifying some of the potential pitfalls in future communications between government and the public.

In early 1980, I visited Three Mile Island to consult with the EPA staff. They had been in Pennsylvania for about one year to monitor radiation levels in the area, following the reactor accident in 1979. During the visit, my EPA colleagues thought I should meet with some of the molders of public opinion in the area.

We went to a church where 25 pastors from the area had gathered at the request of the government officials supervising activities at the reactor site. The idea was to persuade the clergy, and through them a large number of local residents, that radiation was being effectively contained within the disabled reactor and that there was no need for concern. Also, experts would describe the plans for a one-time release from the reactor building into the atmosphere of xenon, a radioactive gas which had been accumulating and thereby complicating efforts to clean up debris from the accident. They would explain how the release would be completely harmless: the radiation levels in surrounding communities would be so minuscule that the most sensitive instruments would have difficulty detecting the xenon. The clergy, therefore, should calm the fears of their parishioners who might imagine that a radioactive cloud was to engulf the region.

This meeting was a lesson in how not to interact with the public. A group of representatives from different federal and state agencies presented a set of briefings on how nuclear reactors function, how the accident occurred, how the xenon would be released, how levels of environmental radiation are measured, and why low levels of radiation have little health and environmental significance. The speakers droned on for 90 minutes, occasionally complaining that the church was not equipped with appropriate projection equipment. The pastors did not intervene with a single question. They twitched with apparent boredom, but they restrained themselves from dozing.

Finally, as the time to adjourn the meeting approached, the senior government spokesman asked if the pastors had any questions. Finally, one minister spoke, "We appreciated your going to all this trouble, but what you say is not important. The issue is not radiation levels, but it is the morality of using nuclear power. We don't need nuclear weapons, and we don't need nuclear reactors." With that statement, the pastors excused themselves and left.

Another case in point came 15 months later as the EPA grappled with the problems at Love Canal. Upon my arrival in Niagara Falls, I was greeted by a police officer who advised me that he would be my escort during my brief stay in the city. He added that bodyguards had been assigned not only to me but to all government officials who had arrived. We were in the city for a meeting with the residents of the Love Canal area concerning the results of the EPA monitoring program which had been carried out to determine the habitability of the area. As discussed in Chapter 2, this program was directed to the residential area beyond the ring of homes immediately adjacent to the Canal. The homes in the inner ring had already been evacuated.

My bodyguard suggested that we have dinner at a small out-of-the-way café since reporters and hostile homeowners seemed to be everywhere. We could then proceed directly to the chambers at city hall for the meeting. At dinner the policeman described the pent-up anger of the Love Canal residents over the squabbling between the state and federal governments as to who was responsible for investigating and cleaning up the mess.

After dinner we joined the other ten officials and their police escorts at city hall exactly on time, and my government colleagues and

I immediately took seats at the front of the council chambers. More than 50 television and press cameras and several dozen police officers ringed the audience of 300 homeowners. The mayor presided.

Using large maps and charts, we carefully explained the design of the monitoring program and the levels of environmental contaminants which had been found throughout the area. We then articulated the bottom line: (1) the levels of contamination in the area where residents still lived were no higher than contamination levels in other industrial cities of the United States; (2) there was no evidence that contamination in the inhabited area was attributable to dumping in Love Canal; and (3) the levels of contamination posed no threat to human health. The meeting was opened for questions, and the shouting soon began.

A small bespectacled man quickly rose and began waving documents. When asked by the mayor to identify himself, he revealed that he was a lawyer speaking on behalf of some of the owners of property where we had conducted the study. He displayed outrage over the government's "whitewash" since "it was obvious" that the entire area had been "poisoned by leakage of chemicals from the dump." As he began to rattle off rebuttal evidence from *his* group of scientific experts that the area was indeed contaminated to an unsafe level due to its proximity to the Canal, the cameras pressed in around him. The police edged forward, and other members of the audience a few feet away began to shout him down.

For a moment the meeting took on a circus atmosphere as the shouting spread throughout the chamber. The television cameras switched back and forth, and the police kept hustling from one position to another. The lawyer's adversaries had much louder voices. They immediately made it clear that the lawyer's clients, while still homeowners, had already moved from the area on the lawyer's advice that the neighborhood would be declared a disaster zone by the government. The residents still living in the area were not interested in a legal treatise presented by a lawyer representing absentee landlords who was trying to squeeze as much money as possible out of Occidental Petroleum Corporation—the parent company of the dumper of the chemicals 30 years earlier. These residents had come to the meeting to learn the "facts" from this latest study, and they were genuinely interested in what the EPA had to say.

The mayor gained a semblance of control over the meeting, but shouting and acrimony continued to punctuate the discussions. Most residents, vitally concerned over the health of their families, seemed relieved by the government's contention that the threat had been exaggerated and that evacuation was not necessary after all. Some were clearly incensed that a lawyer was trying to discredit a study which they found reassuring, and two or three even seemed ready to engage in fistfights with outsiders who were using up valuable time at the meeting for their own selfish purposes.

The residents were remarkably well informed about the health implications of exposures to toxic chemicals. Most seemed to accept the conclusions that contamination levels were below thresholds of concern and that the chemicals in the environment were from nearby industry and natural background sources and not from the Canal. They did not challenge the EPA's scientific findings. What disturbed them most was the lack of vigorous government action to reverse the image of the area as a toxic wasteland. This image heavily devalued their property and took an emotional toll on their families.

Let me now turn to a final story in community relations. During my tenure as the senior EPA official in southern Nevada in the early 1980s, the EPA developed a novel community outreach program in the towns and small settlements downwind from the nation's nuclear weapons test site. The objective was to reassure local residents that they were not being exposed to harmful levels of radiation from weapons tests.

After discussions with community leaders, the EPA in cooperation with the U.S. Department of Energy enlisted about a dozen high school science teachers to operate radiation monitoring stations in their communities surrounding the test site. The government provided the teachers with training in the fundamentals of radiation monitoring, logistics support, and technical advice, as well as small stipends. The teachers were then able to ensure that the stations operated properly. They collected samples and reviewed the measurements that were recorded on their instruments. They signaled the alarm if there was any semblance of radiation. These teachers had the full confidence of their neighbors who became quickly informed about the monitoring activities of the government and the implications for their communities. As the teachers reviewed the radiation measurements, the credibility of "their" data was never questioned.

In support of this effort, two interesting experiences in interacting with the public took place in 1984 in the middle of the Nevada desert about 100 miles north of Las Vegas and just at the edge of the test site. One was spontaneous and unannounced. We simply dropped in on a family at the largest ranch in the area, devoured large pieces of their homemade berry pie, presented them with some recent popular magazines that had just arrived in Las Vegas, and discussed how they could participate in monitoring radiation in the area. The family was delighted to have visitors from the city, and they were clearly satisfied that radiation would not seep into their ranch.

The second event—a public meeting to discuss radiation from the test site—was formally announced in the local papers and advertised by word of mouth in the desert communities. About 25 local residents showed up to meet with several government officials in a lone café which interrupted the barren countryside along an infrequently traveled desert road. The informal EPA presentation describing nuclear testing was delayed by occasional horseplay at the bar but was greatly appreciated. The many beers consumed by all reinforced the stereotype of the EPA officials, at least in southern Nevada, as real people.

In summary, the meeting at Three Mile Island was not successful largely because of a lack of government sensitivity to the orientation and interests of the pastors. The clergymen simply considered technical details a low priority. Their primary concern was the immorality of nuclear energy. Had the government spokesmen at least acknowledged at the outset that views differ over the morality of nuclear power and invited comments, they might then have had at least an outside chance of reaching the pastors with their technical message. At Love Canal the meeting succeeded in providing residents with some understanding of the results of a highly publicized "definitive" study. At the same time, the mere presence of officials from Washington in the city served as a lightning rod for the frustrations of an entire community after years of study and indecision by federal and state agencies. In Nevada, the personal touch worked well in establishing critical awareness for serious, if informal, discussions. Of course the number of residents of the area around the test site is very small. But even when government officials must deal with much larger populations, taking the time to interact informally but seriously with a few key public leaders can have substantial payoff in building mutual confidence and respect.

The Conventional Wisdom for Communicating with the Public

Why doesn't the public trust government agencies responsible for environmental decisions? Is it because of the unacceptable behavior of the EPA leadership during the early 1980s, and specifically their activities that involved collusion with industry, perjury by key officials before the Congress, and a jail sentence for a top official? Is it because the Department of Energy, behind the shield of national security, dumped nuclear wastes at its facilities with insufficient regard to the lasting damage that would ensue if the wastes were not properly handled? Is it because the public won't accept the idea that there must be losers in every environmental decision? Or is it simply a matter of poor communications between the government and the public? Public mistrust is rooted in all of these factors.

Now as the government agencies struggle for a degree of public confidence in their activities, *effective* programs for communicating with the public have become a high priority. Every environmental agency throughout the country has a public relations program. Many of these programs have assumed a refreshingly high level of sophistication. They no longer simply distribute press releases but genuinely seek to engage the public in discussions of issues. Thousands of public meetings concerning environmental issues are organized each year by these agencies. Most importantly, decision officials have become the principal participants in efforts to communicate with the public.

Meanwhile, risk communication has become a trendy topic for academic researchers. What can environmental experts say now concerning the effective communication of risks after two decades of intensive experience in addressing environmental and health problems—other than noting that there is no simple formula for successful communications?

First of all, the American public simply will not accept the naïveté reflected in the following statement which was recently published by specialists in risk communication: "The job of risk communication is . . . to explain risk assessment findings to the public so that the public accepts the judgments of the experts, thus saving the day for rational risk management."[5]

More sensible communications experts note that differences in the ways risks are expressed can have a major impact on public perceptions and personal decisions. For example, if a hazard is described as a risk to a general population (e.g., the incidence of cancer in the area will increase by one tenth of one percent), it is much less worrisome to individuals than if it is described in terms of personal impact on those individuals (e.g., your children are in a higher risk category than children living in other areas). Also, when people are informed about an unfamiliar hazard, say a pesticide with a complicated chemical name, they often will generalize to other hazards, such as other better-known pesticides or other chemicals used around the house. Risks linked to dreaded diseases such as cancer and birth deformities are generally perceived as more threatening than risks which might have less familiar effects such as nervous disorders.[6]

Two of the most helpful documents in introducing both specialists and laypersons to the science of risk communications as this neglected field moves from the backwater of cognitive psychology to the center stage of public policy were recently published by the EPA and by the American Chemical Society. These brochures have been distributed to thousands of local public health officials and local community leaders, and they have become very popular in schools as well.[7]

The pamphlets recognize the difficulties that arise as communities and individuals try to understand the significance of industrial releases of chemicals into nearby communities—releases that can excite fears and anxieties regardless of the extent of the risk to human health. The pamphlets note that government officials must frequently face public outrage or total apathy concerning local problems. Citizens often demand absolute answers. In the absence of very persuasive data, the public will be reluctant to change strongly held views. Government agencies often have difficulty communicating with the public due to their own limited understanding of the interests, concerns, fears, priorities, preferences, and values of individual citizens and public groups.

Given these realities, the documents offer the following guidelines for government officials responsible for communicating with the public about chemical risks at the local level:

- □ Accept and involve as legitimate partners all parties with an interest or stake in the issue. The goal should never be to manipulate the public into accepting decisions that have already been made or to justify avoidance of action. Adequate time should be set aside for hearing concerns, generating alternatives and solutions, and making decisions.
- □ Listen to the audience. Assumptions should not be made as to what is bothering people. Sometimes people are more interested in the social and psychological dimensions of risks (e.g., voluntariness and controllability) and in the trustworthiness, competence, and credibility of industry officials and community leaders rather than in the scientific intricacies of the problems.
- □ Assess and nurture government credibility. Credibility is difficult to earn, easy to lose, and, once lost, almost impossible to regain. It is of great importance that concrete actions follow promising words.
- □ Plan carefully before communicating. There are many publics, each with its own interests, needs, concerns, priorities, perceptions, and preferences. Communications should be targeted to specific audiences.
- □ Be honest, frank, and open. The strengths, limitations, and uncertainties in data and in assumptions, including judgments of other credible sources, should be discussed openly. Mistakes should be acknowledged.
- □ Speak clearly and with compassion. Technical language and jargon pose substantial barriers in communicating with the public. Of particular importance, the qualitative dimensions of risks such as equity and fairness should be acknowledged.
- □ Coordinate with other credible sources. Third-party experts can be very useful in enhancing perceptions of objectivity. Conflicting interpretations are usually unavoidable and may help illuminate different points of view.
- □ Meet the needs of the news media. The media is often more

interested in simplicity than in complexity. Officials should take special pains to be accessible to the media and should respect the space and time constraints of the media.

After living through the EPA scandals of the early 1980s, environmentalists and indeed many other segments of the public were heartened when in 1985 they read the following instructions from the new administrator of the EPA to his employees:

> Most important is the integrity of the EPA; we must do business in the open. The nation is interested in the values we reflect in our work—in how we work as well as what we do. EPA managers should actively solicit and respond to the advice of interested parties, such as states, industries, environmental groups, and the general public . . . . Our proposed actions to improve environmental quality must be explained in ways that encourage people to suggest changes that may make our actions conform more closely to public values.[8]

Yes, a few ground rules for communication between government agencies and the public have been set forth. Now the challenge is for all interested parties to respect these guidelines in ensuring that diverse points of view are presented in a timely manner and are fairly considered.

Withholding Scientific Data from the Public

A particularly difficult issue confronting environmental agencies is when to release scientific data to the public that suggest the possibility of a toxic hazard. New information showing adverse biological effects may be developed during laboratory or epidemiological studies. Scientists may obtain monitoring measurements which indicate high levels of chemical pollutants in the environment. They may uncover new evidence of excessive discharges of chemicals from manufacturing facilities.

If a scientific study is still under way, should data being analyzed during the course of the study be released before the study is completed? Would withholding the data jeopardize the health of residents who would leave an area if they were aware of a possible hazard? Would piecemeal release of the data suggest a degree of hazard or a level of complacency that might prove erroneous upon completion of the study?

Let me cite an example. In 1975 the EPA began sampling rivers and streams throughout the country for the presence of asbestos. During preliminary testing of the Schuylkill River which fed the Philadelphia drinking water system, the EPA found asbestos fibers in one of two samples taken from the river above the drinking water intake. Obviously, more extensive sampling was needed. However, only one laboratory in Chicago was equipped to analyze water samples for low levels of asbestos, and a delay of several weeks could be expected before a significant number of samples could be analyzed.

In the meantime, EPA specialists postulated that asbestos fibers were released from manufacturers in clumps and transported in mini-packets through the discharge pipes and then downstream. Thus, they concluded, the sample may have been a "fluke" that was typical of the overall conditions of the river. Still, the EPA had found asbestos, a hazardous material, in the river, and the Agency felt obliged to inform local officials. The city immediately released the information to the press, and headlines soon blared the threat of asbestos fibers in the city's water supply. Subsequent monitoring confirmed the initial hypothesis that the samples were indeed atypical and that asbestos was not present to any significant extent in the water supply. A false alarm which greatly exaggerated the environmental threat was sounded to the public. Some residents undoubtedly shifted to bottled water on the basis of the press reports. Fortunately, the city environmental experts who did not want to release the data in the first place successfully argued with the city politicians against any action to tamper with the water supply until the results of additional monitoring had become available.

In the environmental field, scientists never have "adequate" data. Environmental officials have become accustomed to acting on the basis of limited data. The public too must frequently cope with limited information. But should data be released before the accuracy of the measurements has been confirmed? Before confirmatory data have even been collected? Before scientists have had an opportunity to review the data for inconsistencies, trends, and patterns? Before scientists and decision officials have assessed the significance of the data?

While each case is different, political factors almost always affect the timing of data releases. In some cases, policy officials and other interested parties may agree well in advance on a specific date for a public release, and scientists try their best to adjust their timetables to meet this deadline. Often an environmental agency asks, "Do we have enough information for a press release that will not unduly excite the interested parties but will satisfy them until the study is completed?" Finally, the press may become aware of preliminary findings through its own channels, leaving the government with no choice but to release the available data. The approach of simply holding back information until an agency has finished its studies is very difficult to support in the face of political pressures. Also, a myriad of ethical considerations cloud almost every decision when public health is at stake.

Given the scientific uncertainty associated with partial data but also recognizing the right of the public to know promptly about the results of government investigations, I have several suggestions concerning the release of information obtained by the government on environmental hazards. First, data which have not been validated for their accuracy through duplicative measurements or other cross-checks should not be released, even with caveats. Agencies should not attempt to interpret data which could be biased by poor collection or laboratory practices, nor should they expect the public to wrestle with data distortions. If the press obtains unvalidated data, the government should simply respond that at that time no comments can be offered on questionable data. Second, whenever an agency releases partial data concerning an environmental problem, the agency should explain how far along the overall study has progressed and should present its preliminary views, if any, on the significance of the partial data. Finally, if there is a question as to whether to release *validated* data, the agency should tilt toward releasing rather than withholding the data.

Industry too is often saddled with conflicting impulses on when to release data to the public. The public relations departments of companies are usually eager to provide nearby residents with accurate information about the operations of their plants. They want to nip in the bud any false allegations about environmental contamination resulting from manufacturing processes. On the other hand, corporate legal staffs argue for releasing information only when required by law or regulations. Any additional information might unexpectedly be used in legal actions against the company, they contend. Indeed, corporate lawyers worry about releasing information even to plant workers who in turn might disperse information to others or who might use the information in claims for workers' compensation.

Meanwhile, industry is required to report to government agencies information on spills of chemicals and on other hazards posed to workers by their activities. Also, since the mid-1980s federal regulations call for industry to provide annual estimates to the EPA on a plant-by-plant basis of the amounts of several hundred chemicals being discharged into the air and water or being sent to waste storage sites. The EPA in turn makes this information available to the public through press releases, through reports containing compilations of data that are submitted, and through computer printouts of the raw data.[6]

The community's right to know about chemical activities is now embedded in other regulations as well. Industry is required to provide the EPA with any evidence it uncovers suggesting that chemicals which it manufactures might be hazardous to health or the environment. For example, a company may at its own initiative conduct a study of the effects of a chemical on laboratory animals. If the results of the study suggest that the chemical might induce adverse reactions, the company now must promptly inform the EPA—a requirement triggered by industry's withholding of laboratory studies indicating harmful effects of exposure to vinyl chloride as discussed in Chapter 2. The Agency then decides whether the information should be widely disseminated. Also, companies must affix appropriate labels on chemical containers setting forth any hazards associated with the chemicals.

Cutting across these obligations of industry to report on their activities to government agencies are the requirements of the Freedom of Information Act. This law provides for public access, on demand, to information which the government possesses. At the same time, however, industry is guaranteed protection of its industrial secrets. Both government and industry devote considerable effort to devising techniques for complying with these two seemingly contradictory requirements. For example, a common approach to protecting the secretive molecular formula of a chemical is for the company to identify for public release the class of chemicals within which the protected chemical falls. Generally, chemicals in the same classes tend to have similar hazards, and therefore the public is warned about the potential hazard of a class of chemicals. This approach is not completely sound since even very similar chemicals behave differently, but it still provides a reasonably good indication of hazard.

Mediation Tempers Confrontation

The generators of pollution, environmental groups, and other segments of the general public all want to be informed of governmental studies and regulatory actions. However, each frequently wants to exert influence on these activities as well. Each can, of course, participate in the highly structured legal proceedings for "public input" established under many laws. But those outside the government often feel that such participation is little more than perfunctory and cannot truly influence large governmental organizations.

For more than 15 years, a few environmental organizations in Washington and in other regions of the country have searched for effective approaches to speedily resolve environmental problems among the key interested parties themselves, and particularly those types of problems that frequently become entangled in a time warp of legal proceedings. Many environmental activists and industrialists as well have simply lost patience with the high administrative and legal costs in resolving contentious environmental stalemates. The costs of formal hearings, written rebuttals, and court rulings that drag on for years have become legendary.

Environmental mediation can be an attractive alternative. It can encourage the public to participate in debates over environmental disputes while reducing the factors of time and cost. This approach was conceived as analogous to labor–management bargaining under the auspices of an impartial mediator. Participants at the table include representatives from industrial, municipal, or other organizations whose actions may disrupt the environment and representatives from the public segments who would be most closely affected by the environmental impacts. The mediator may be a university professor, a respected local figure, or someone else who is perceived by all parties as objective.

The idea is to reach a consensus on some or all aspects of a specific environmental problem during relatively unconstrained discussions among the key parties. Such discussions, free of the formality of a judicial setting, can lead to agreements which are immediately implemented by the participating parties or alternatively presented to a regulatory agency or to a judicial body for legal codification. Early environmental mediation efforts were directed to issues of land use such as the siting of dams, the zoning of real estate, and the routing of high-tension lines. Then in the late 1970s, representatives of labor and industry attempted to close the gap between these two interest groups over the regulation of carcinogens in the workplace, but this topic was soon perceived as too important to be left to informal mediation efforts. More recently industrial and environmental groups have successfully worked out some of the procedures to be followed by the EPA in requiring industrial testing of chemicals under the Toxic Substances Control Act.

One highly publicized mediation effort helped resolve a landfill siting dispute in East Troy, Wisconsin, in the late 1980s. Under state law, the state agencies were charged with determining the need for the landfill. Despite the protests of local citizens, the state ruled in favor of the landfill and supported the engineering soundness of the proposed facility. However, the state determined that the town was to have a say on the "economic and social" dimensions of the landfill.

In the formal negotiations, six local citizens, supported by a lawyer, represented the town and county while the owner of the proposed landfill also hired a lawyer. For a year, negotiations remained on dead center which was fine with the townspeople who preferred that the landfill never arrive. Then, the landfill owner hired a new lawyer from a more powerful legal firm, and the dress code in the negotiations changed from flannel shirts to three-piece suits. The owner decided to demonstrate to the local waste-siting board that the townspeople were not negotiating in good faith and therefore local approval was no longer necessary. The board disagreed. Meanwhile, the local citizens in turn charged the owner with not acting in good faith since the new lawyer withdrew all previous informal agreements.

Finally, after three years of haggling, the parties agreed to engage a mediator. Within a few months, the mediator succeeded in developing an acceptable 41-page agreement that became binding on both parties and enforceable in the courts. A key issue was the monetary compensation to be paid by the owner to the town. The landfill soon began to operate.[9]

In this case mediation brought a long and contentious process to an end in a reasonably short time. It provided the community with some assurance that local interests had been protected. If the community must have a second landfill, monetary compensation and stringent environmental safeguards beyond those required by state law were some consolation.

Environmental mediation remains an attractive and underutilized alternative to the more formal procedures for regulating chemicals. However, the public official who is ultimately responsible for ensuring that the environmental hazards under discussion are controlled must be fully committed to mediation as a serious effort. In effect, a governor or an EPA regional administrator, for example, must delegate some of his or her authority to the mediation process and must be prepared to accept the outcome.

One of the most difficult issues in structuring environmental mediation deliberations is who is to represent each side, particularly the public. Can two or three local or even regional organizations adequately speak for a broad public? A second issue relates to the continued mistrust between environmental groups and industry. If industry makes concessions during mediation efforts, will the environmentalists nonetheless resort to the courts for additional concessions after mediation ends? Conversely, is industry to be trusted in implementing everything it promises?

Chemical problems confined to small geographical areas lend themselves more readily to informal resolution than those which permeate the entire country. If only one or a few manufacturers are involved, and if there is good faith on the part of industry and the local residents, satisfactory solutions to specific problems can often be worked out.

A second type of bargaining that has gained wide acceptance is the *consent decree* which is now used routinely as an alternative to lengthy court proceedings for resolving contentious issues. Under this procedure a judge oversees negotiations between the interested parties outside the courtroom to come to agreement on steps to resolve an environmental dispute. For example, they may come to agreement on acceptable levels of pollutants draining into a specific stream.

At the national level, the National Resources Defense Council frequently takes legal action through the courts to force the EPA to initiate or strengthen specific regulations. Sometimes other organizations also join as parties to these legal actions. These parties then work out an approach with the EPA which they all deem acceptable, operating under court-mandated deadlines. They eventually sign a consent decree which is approved by the judge and has legal standing. This environmental version of plea bargaining is also mirrored in some states to resolve contentious issues.

Finally, in another variation of environmental bargaining the EPA and state agencies negotiate daily with industry, government facilities, and municipalities on the details of many types of permits for discharging chemicals. Also, in Washington EPA specialists are in a continuous dialogue with manufacturers of pesticides and industrial chemicals over requirements for laboratory testing and about limitations that should be placed on the manufacture and use of these chemicals.

Environmental groups bemoan the lack of public involvement in these types of discussions. In my view, the current system which calls upon the EPA to represent the public interest in such detailed negotiations within boundary conditions set forth by law is quite appropriate. There is no need to further complicate the already complicated negotiations through greater involvement of the public in these very detailed and often very technical discussions. Indeed, regular congressional reviews of these activities should provide the necessary watchdog function for the public.

Conducting the Business of Government in a Glass House

In general, the EPA and most state environmental agencies have been on the right track in their efforts to involve a broad segment of the public in developing approaches to reducing risks from toxic chemicals without becoming bogged down in so many details that action is indefinitely delayed. The agencies are continuously improving their approaches to respond to the oft-repeated adage, "The challenge is to ensure that everyone is in on the action and to still have action."

At its very inception, the EPA prided itself that it would be a glass house open to all. After an unfortunate detour in the early 1980s when the Agency's leaders decided that they alone should chart the nation's future environmental course, the EPA has returned to its initial tack of reaching out for suggestions from all quarters. This open attitude has been strikingly different from the approach of the Department of Energy, for example, which until 1989 suppressed its environmental transgressions behind the mystical veil of national security. Now, as the Department of Energy and many other federal and state agencies try to open their deliberations to greater public scrutiny, they can learn a great deal from the EPA's record.

At the same time, EPA employees have been frustrated by the distortion of their own priorities resulting from heavy public involvement in the business of government. In 1987 a group of senior EPA career officials examined the impact of the Agency's programs on reducing all types of environmental risks and reached the following conclusion: "Overall, the EPA's priorities appeared to be more closely aligned with public opinion, often expressed through congressional mandates, than with estimated risk." These environmental professionals concluded, for example, that the Agency's high-priority programs to clean up hazardous waste were less important in reducing environmental risks than its lower priority programs directed to the control of pesticides and the reduction of indoor air contamination.[10] While some skeptics in the Congress and in the academic community take issue with the list of priorities developed by these EPA officials, none denies the importance of public perceptions and public pressures in setting the regulatory agenda.

Despite the commitment to reach out for meaningful input from the public, the EPA's efforts to involve interested parties in its regulatory activities are often in conflict with the realities of bureaucratic life. This situation is particularly evident in the development of national regulations. The EPA expends great effort, involving dozens and sometimes more than 100 specialists from the Agency and from other agencies, in the development of a major regulation. Consensus building is the order of the day, and the personal agreements that are reached within the Agency and among agencies are frequently tenuous at best. Thus, by the time a regulation is proposed for public review and comment, the EPA authors have a pretty firm opinion as to how the final regulation should read if it is to continue to command support throughout the bureaucracy regardless of the public comments received. Often in their view, changes proposed by the public simply complicate the task of retaining the consensus within government.

I have participated in public meetings on new regulations, both as an EPA proponent of regulations and on other occasions as a member of the interested public making suggestions for modifying regulations. At these hearings government officials dutifully record all comments and usually respond to questions. But these regulators usually are not interested in performing major surgery on proposed regulations regardless of the merits of the arguments. They already have devoted two to three years developing the regulation, and they have one overriding objective—namely, to publish the final regulation as soon as possible.

A different approach is necessary to avoid expediency at this point and to ensure that public comments are taken more seriously by the EPA as well as by state and local officials who prepare regulations. Specifically, the EPA official who is responsible for the development of a proposed regulation should not be responsible for the review of public comments on the regulation or for the preparation of the final regulations. He or she is simply too wedded to every nuance of the original proposal. Another official not involved in preparing the proposed regulation should be given the job of reviewing comments and shepherding the final regulation through the bureaucracy. This is contrary to current practice which usually calls for the same official to handle a proposed regulation over all the hurdles from inception to final promulgation.

Meanwhile, as previously discussed, debates of risk between government agencies and the public should not be confined just to hearings on regulations. They should not be viewed as special events orchestrated by specialists in public affairs. They should become a routine dialogue and an integral part of the job of every regulatory official.

In order to save time in meetings with the public, government officials often resort to films, slides, and other slick presentation material. However, this approach may be inviting negative reactions. I remember attending a public meeting organized by the Department of Energy in Henderson, Nevada, on nuclear waste. As soon as a film projector was set up, the crowd groaned and one attendee could be heard by all saying "another snow job." Too often the projector smacks of an elementary school setting, and the government is perceived in a patronizing role as the teacher with the public as the pupil.

In general, scientists are poorly prepared to interact effectively with the public. They usually try to be so precise that no one can understand them. Who cares other than lawyers whether the concentration of a chemical in the soil is 5 ppb or 10 ppb? Who understands what is meant by a picocurie of radiation? Scientists may comprehend what happens to animals in laboratory experiments, but in the eyes of the public the scientists really don't look at human health risk the way the family doctor does. When the discussion turns to the uncertainty of risk, the dialogue becomes totally incomprehensible.

To some environmental specialists, the public is obsessed with the notion that minuscule levels of trace chemicals cause cancer or birth defects. To others, the government is callous to the contention that miscarriages, asthma, and cancer are being caused by the by-products of industrial processes.

The media will continue to play a crucial role in shaping public attitudes. While reporters may be paid simply to report, in the environmental field they cannot be dismissed as only messengers between government and the public. They have personal views on the issues at hand which often affect themselves as well as others. They must be recognized as important participants in the process of determining the societal response to risks.

However, newspapers and television are not substitutes for more intensive environmental education. Scientists must be retooled to communicate in simple, understandable language. The public must become more sophisticated in its appreciation of the risks and benefits of chemicals. We all need to become more sensitive to the interests and concerns of the many parties vying for a safer environment. Our nation simply cannot afford to be blown by the winds of public emotion lest social paralysis inhibit our advancement as an industrialized nation.

5 Cleaning Up the Wastes of an Industrial Economy

Take your refuse elsewhere or you will be fined.
—A Roman signpost

Waste: Any unlawful act or omission of duty on the part of the tenant which results in permanent injury to the inheritance.
—Black's Law Dictionary

Cleanup: Actions taken to deal with a release or threatened release of hazardous substances that could affect health and/or the environment.
—U.S. Environmental Protection Agency

Toxic Wastes Invade Every City and County

After returning from Love Canal to the Las Vegas laboratory in 1981, I was both relieved and depressed. The EPA had determined that at least for the short term the wastes in the Canal posed no threat to the residents still living in the area. The wastes that were present could be contained within the Canal without difficulty for the next few years.

However, the containment solution of the EPA relied on an underground drainage system around the two-mile perimeter of the Canal. The drains channeled any leaking liquids into catchment basins. The chemicals were then removed and sent to a nearby disposal facility. In the long run such an approach would surely be exorbitantly expensive. Millions of dollars were required each year to operate the system and to monitor the condition of the Canal with little assurance that the system would not break down or that such a Band-Aid solution could hold political support in the future.

At that time, estimates by engineering companies of the cost of trucking out the wastes were in the range of 400 to 800 million dollars (estimates which seemed then and now to be much too low). These estimates assumed the wastes could be removed without excessive risks of explosions or releases of dangerous chemicals during the extraction operations. The EPA quickly rejected this solution for removing the wastes on financial grounds alone even though in the long run a one-time high cost would probably be lower than the cost of superintending the leaking wastes over the next one or two centuries.

Thus, the soundness of the approach of containment and controlled drainage seemed questionable at best. But in view of near-term realities as to cleanup funds which were available, I had no better alternative solution. In any event, specialists from the Las Vegas laboratory had completed their assessment of the problem, and future responsibility for Love Canal within the Agency resided with the EPA regional office in New York City. Thus, our laboratory specialists turned their attention to other waste sites.

While I had been concentrating on Love Canal, the laboratory's photographic interpreters had already been hard at work scanning aerial photographs of a number of metropolitan areas. They had discovered over 100 locations on Staten Island alone, for example, which looked like abandoned waste sites. They had found hundreds of suspicious indicators of refuse on photographs of Pennsylvania, and the photographs of Virginia had been so heavily annotated by the interpreters that we could hardly recognize the Commonwealth. Clearly, prior to these photographs the EPA had been unaware of hundreds of these potential problem sites around the country. A modest effort using existing aerial photographs netted an unexpectedly large number of questionable sites.

In a related effort, the laboratory had provided support to a small environmental office in a county in upstate New York which had undertaken an inventory of abandoned wastes within its jurisdiction. In addition to using aerial photographs, the county environmental specialists had enlisted the local population in a hunt-the-dump campaign. Based on telephone tips from concerned residents, the county officials had located dozens of sites where dumpers had abandoned barrels, boxes, and loose sludges laden with chemicals.

Armed with aerial photographs and reports from the county in New York, I flew to the EPA headquarters in Washington confident that the Agency could mount a low-cost nationwide campaign which would provide an authoritative assessment of the extent of the abandoned waste problems. The Congress had recently enacted the Superfund law, and an obvious first step seemed to be to size up the problem.

The reaction at the EPA headquarters to my overtures for a coast-to-coast hunt-the-dump program was discouraging. "The Reagan Administration doesn't want to hear about problems. It wants to hear solutions." "There are enough well-known Superfund sites to keep the EPA busy for a decade. Let's not identify more and simply complicate the task." "The Las Vegas laboratory should direct its efforts to supporting cleanup actions that are under way and should forget about trying to change the Agency's priorities." "Deep-six those photographs, and forget about the project in New York." We complied with these instructions, except we saved the photographs.

Three years later, in a total about-face under pressure from the Congress, the EPA headquarters instructed the Agency's regional offices, with support to be provided by the Las Vegas laboratory, to identify all abandoned waste sites that required remedial action across the nation, and our specialists dusted off their photographs. In 1990, EPA had 32,000 sites in its inventory of potentially hazardous sites, with preliminary assessments indicating that further action was not necessary on 14,000 sites.[1]

Meanwhile, as the laboratory complied with the original instructions of 1982 to concentrate on well-known sites, I undertook a three-week tour of a few of these toxic dumping grounds to gain a better appreciation of the on-site problems. The laboratory was particularly interested in providing field-monitoring crews with portable equipment that could be used to determine the presence of dangerous chemicals that were escaping into nearby residential areas or that could pose a threat to the field crews themselves. For example, laboratory scientists were evaluating hand-held instruments which NASA had developed for sensing the presence of toxic metals on other planets. Also, they were testing the accuracy and reliability of monitoring devices which were small enough to be carried in shirt pockets. Occupational safety specialists had developed these devices for monitoring air quality in chemical plants. Finally, our scientists had purchased an array of portable geophysical instruments for probing the conditions under the surface of the Earth in their search for buried barrels and for plumes of chemical contaminants in shallow groundwater.

My first stop was at a group of abandoned sludge ponds on the edge of Tacoma, Washington, where I witnessed negotiations between EPA lawyers and lawyers for the company responsible for the chemical gunk. The company accepted its responsibility for the cleanup, and the lawyers debated the location of a well for monitoring the condition of the groundwater. Meanwhile, a rig mounted on a large truck stood by ready to move 30 feet to the left or right to drill the well once the negotiations were concluded. However, an agreement could not be reached. The lawyers instructed the drilling crew to wait for several days at a cost to the company of $2000 per day until the experts could meet in Seattle and come to a conclusion.

At the Springfellow site near Riverside, California, the enormous amount of industrial waste which had been dumped into a very large canyon for many years was an overpowering sight. We stood atop a clay barrier that went down 90 feet to the floor at a narrow point in the canyon. The waste was packed in from the barrier back up the canyon for 200 to 300 yards. The only problem was that the barrier looked good at the surface but was full of leaks under the ground. Heavy metals and organic solvents had migrated through the barrier into water wells several miles below the barrier. Inaccurate press accounts fueled rumors that the drinking water was contaminated even though the wells in the area were used primarily for agriculture and had not provided drinking water. At that time, after years of investigations of the problems at the site which had already cost tens of millions of dollars, the EPA would again provide $1.5 million for still another study to determine what needed to be done.

Near Galveston, Texas, I visited a Superfund site along a very busy freeway. The contractor hired by the EPA to assess the condition of the site and to begin cleanup was proceeding very slowly due to safety concerns. Each field worker entering the site spent 40% of his or her time donning or removing protective clothing and passing through hygienic checkpoints. The safety personnel at these checkpoints outnumbered the field workers. This procedure, which had become standard in the region, seemed a little excessive since the preliminary assessment did not reflect such a high degree of hazard as to warrant these extreme precautions. Meanwhile, passersby walked within several feet of the fence erected but 10 yards from the contaminated areas, and monitoring measurements did not indicate any problems.

At Woburn, Massachusetts, our jeep drove over hundreds of acres of buried wastes. Several chemical companies had accepted partial responsibility for the wastes, and they had engaged engineering firms which were busily assessing the extent of groundwater contamination. Many rigs for drilling monitoring wells were on-site. The EPA project officer was convinced that contaminants would follow the drills down the shafts of the wells but was reluctant to intervene lest he be accused of delaying progress. Also, several engineers noted the inadequacy, due to financial limitations, of the underground venting system. It had been installed a few years earlier to allow gases to escape through the soil and thereby reduce the possibility of buildups of pockets of dangerous gases which might explode. Residents of the area appeared numbed by the whole experience. They had given up on the likelihood that the EPA would come to their rescue. They didn't trust the assessments being directed by the responsible parties under what they considered the less-than-watchful eye of the EPA.

The visits to 15 sites left me with several impressions which paralleled the conclusions of hundreds of other case studies of waste areas which I had read in the popular press and in the scientific literature. First, each site is unique, and approaches to assessing and cleaning up the site must be customized to its specific characteristics and to the local political as well as environmental conditions. Also, the scientific and engineering aspects are often highly complicated. Unfortunately, too often these very complexities are used as an excuse for inaction. Finally, though cleanups are expensive, shortcuts to save money usually end up increasing costs in the long run. Our nation simply cannot afford return visits to sites to correct the shortcomings in initial assessments or cleanup actions.

Turning to the field operations, I was repeatedly struck by the dedication but the lack of experience of the EPA employees who were assigned as project officers for the sites. The specialists from the state environmental agencies working at the sites were somewhat more seasoned, but they were primarily classical geologists who were entering unfamiliar territory of chemical pollution. The de facto management of the site activities was in the hands of contractors, and they relied heavily on young specialists who were in their first professional positions following recent graduation from college. Somewhat belatedly, in 1990 the EPA began assigning senior scientists to assist government project officers at sites throughout the country, and this influx of technical expertise should help the Agency gain greater control over cleanup details.

More important, however, was the lack of decisiveness in the cleanup approaches at the sites. Nowhere did I encounter a passionate determination among the contractor specialists, in particular, to clean up the wastes as rapidly as possible and either restore the area to a usable condition or, at a minimum, ensure that the surrounding areas would be free of a dangerous eyesore. The contractors were methodically following instructions of the EPA or the state agencies, and only infrequently did they demonstrate the type of initiative and aggressiveness needed to overcome the long list of obstacles for an effective cleanup. In several cases, the approach was simply to "stabilize" the wastes (an undefined concept at best) and then to pave over the site or plant sod on contaminated areas so they would look good even though many specialists on the scene scoffed at this solution.

When asked about "permanent solutions," field personnel usually dodged the question and complained about the "system"—a system controlled by officials in Washington, the state capitals, and the EPA regional offices. These officials, they believed, were seeking as many quick fixes as possible. Many would soon go on to other pursuits, and most of them were so tied to their desks that they were out of touch with the realities of cleaning up the environment anyway, according to the workers in the field. Such comments were not entirely fair since senior environmental officials were wrestling with a new, politically charged program which was being constantly changed by shifts in congressional attitudes and new budget priorities.

Cleaning up the trash of our predecessors is a dirty job. It is time-consuming and expensive. It is anything but glamorous, but it must be done. Corrective actions have "stabilized" many waste sites, at least for the present, but many others are still leaking. Determined and relentless cleanups can greatly reduce those leakages which threaten our natural resources.

Most importantly, attitudes among a number of leaders of the industries largely responsible for the waste problems have changed significantly during the eight years since these visits from recalcitrant, insensitive, and calloused to an apparent determination to work with the government to solve the problems as rapidly as possible. One highly visible indication of the commitment of some large companies to cleanups has been the establishment of Clean Sites, Inc., a corporation formed by industry and environmentalists to help clean up sites and to mediate site negotiations. Hopefully, this attitude of increased industrial responsibility and cooperation together with a greater readiness to share cleanup costs will permeate all industrial organization in the years ahead.

Expectations and Disappointments of Superfund

President Bush's commitment to the Superfund program seems clear enough: "I'm for an aggressive, no-nonsense approach to cleaning up toxic waste dumps. I'm for strengthening enforcement against dumpers, quickening the pace of our cleanups, and streamlining the bureaucracy that sometimes slows them down."[2]

Recent critiques of the program by congressional organizations, private groups, and the EPA itself suggest that the political, financial, and technical problems surrounding the program will not be easily solved. President Bush will have left office long before we have achieved sufficient progress to warrant optimism that we as a nation can eventually stop the flow of chemicals into groundwater and the atmosphere.

Still, the president's leadership in the immediate future is essential if the program is to move forward at a rate that will allow the nation to conquer the problems of hazardous wastes by the end of the century. He needs to strongly support budget requests for environmental protection. He should take the lead in stimulating greater contributions to cleanups by state and local governments. He must become the point man in forging a better partnership between the Congress and the executive branch—a partnership that helps the EPA withstand local political pressures for funds for favorite projects and that convinces the Congress to remove some of the legislative barnacles currently shackling administration of the program.

In its fourth stinging critique of Superfund in 1989, the Congressional Office of Technology Assessment called for scrapping some of the program's basic approaches. The report was particularly critical of the way that the EPA pushed more of the cleanup responsibility onto the industrial polluters who are culpable for many of the waste sites around the country. According to the report, in exchange for greater willingness on the part of the responsible parties to move forward with cleanups, the Agency often settles for less stringent cleanup remedies than would be applied if federal funds were used. The implications of this criticism become clear when considering that industry is expected to shoulder the bulk of the expenditures for cleanups, expenditures which Congress estimates will reach $500 billion over the next 50 years to clean up 9000 bad sites.[3] The EPA denies the allegation that cleanups by industry, under EPA scrutiny, are less thorough than government cleanups, pointing to recent internal studies that document the adequacy of industrial responses.[4]

Meanwhile, the Rand Corporation notes that by 1989 only 18 sites had been declared "clean" despite expenditures of more than $2 billion of federal funds. Rand traces much of the delay in cleanups to the original congressional decision to establish a system that relies in the first instance on forcing industry to pay for cleanups rather than simply having the government pay the bill and, in effect, run a public works program. Rand points out that the current approach results in a series of procedural obstacles related to negotiations with the responsible parties which inhibit prompt action.[5]

While many experts disagree with the implied suggestion of Rand to shift more of the financial burden to the government, the small number of cleaned up sites has been a particularly contentious issue for a number of years and has been a principal reason why the Congress repeatedly places unrealistic deadlines before the EPA in carrying out the Superfund program. The EPA argues that the number of completely cleaned up sites is a misleading indicator of progress since hundreds are in various stages of cleanup, all known near-term health risks have been abated, and complete remediation of a site takes many years of monitoring to demonstrate that there is no residual leakage. However, the EPA's reasoning is politically flawed since the public simply cannot understand why after a decade of effort only a handful of sites can be listed as completely safe. Therefore, the Agency needs to give greater attention to carrying cleanups to conclusion even if cleanup actions are limited to fewer sites.

The amount, character, and sources of hazardous wastes are of considerable importance. In brief, "hazardous" wastes are produced at the rate of 250 million tons per year—enough to fill the New Orleans Superdome 1500 times over. However, this is only 6% of the six billion tons of the total wastes generated each year in the United States. The other 94% consists mainly of agricultural and mining wastes, with much smaller amounts of municipal and public utility wastes.[6]

The chemical and petroleum industries created much of the hazardous wastes found in Superfund sites. Some sites were once municipal landfills that accumulated excessively large quantities of pesticides, cleaning solvents, and other hazardous products which were mixed in with household trash. A few sites are the chemical debris from transportation accidents. Others are the resting places of persistent toxic pollutants contained in industrial wastewater discharges. An increasing number are governmental facilities which in earlier years had been allowed to operate beyond the scrutiny of environmental officials for national security reasons.

What should be done about these wastes? Surely the EPA's longstanding first priority to eliminate any immediate hazards to people or the environment is correct. This priority may require physical removal of the most threatening wastes which could lead to fires, explosions, or other disastrous situations at the sites. Frequently, cleaning up or containing leaking barrels, loose solids, or uncontained liquids which can be washed by rain or melting snow into surrounding areas demands prompt attention. It may also be necessary to capture powdered waste which can be blown about or to interdict underground chemical plumes which are approaching drinking water or food supplies.

Beyond the consensus on taking care of these emergency situations, opinions differ on how to address the longer-term problems of cleaning up sites that blight the countryside and that will eventually become more threatening problems in the future. Constraining the debate are the complexities of the scientific and technical issues. Also, as the EPA has pointed out, the trust fund is finite (currently $8.5 billion) and at present the pipeline is full.[6]

The EPA readily acknowledges that it has neither the financial nor management capability to adequately cope with all of the currently designated Superfund sites, let alone take on additional sites. The present estimate of the average lead time of 13 years that will be necessary between identification of a Superfund site and initiation of cleanup activities may become longer rather than shorter given these limitations.

Critics of the program offer all types of suggestions for its improvement. These suggestions fall largely into four categories: improving the administration and management of the program; deciding when and how to seek cleanup funds from the polluters and when to use the trust fund; establishing priorities for cleanup actions; and determining when a site is clean enough. Given the large sums of money involved, political and economic groups throughout the country will continue to exert pressures to direct larger shares of federal funds in their directions. Their views of administrative efficiency, use of the trust fund, priorities, and cleanliness will seldom be based on objective criteria.

First, with regard to management efficiency and administrative timetables, the Congress should move away from trying to define the details of how the program should be managed by the EPA. Within the constraints already imposed by the Congress, and particularly the ground rules concerning the role of the private sector in the programs, the EPA has been quite responsible in setting in place a reasonably effective program. Congress initially believed that the program would be short term and, therefore, called for heavy reliance on contractors. Indeed, governmentwide policy in almost all areas has a long history of requiring a contract with private industry for a job whenever possible. When Federal Procurement Regulations require the use of the least expensive qualified contractor, the EPA must frequently select less than the best. Also, as noted, when the law requires the Agency to resort to the trust fund only if responsible parties cannot be forced to shoulder the financial burden, the EPA must spend considerable time and effort trying to extract commitments from industry while delaying its own activities.

Auditors unleashed by the Congress are constantly reviewing Superfund activities, and the EPA is required to divert the time of senior personnel to ensure that the auditors receive complete and authoritative information. Meanwhile, the current level of direct congressional involvement in details is already excessive and simply adds unnecessary distractions and complications to the EPA's task. Periodic reviews by the Congress of the EPA's administrative efficiency are important to keep the Agency on its toes, but placing unrealistic administrative timetables and other detailed management criteria in the law is counterproductive.

However, before the Congress will relax the administrative requirements imposed on the EPA, the Agency will have to restore confidence on Capitol Hill that even in the absence of congressionally mandated requirements, the EPA will effectively carry out the intent of the law. In particular, the EPA must demonstrate visible results from its own initiatives. Such confidence building will be difficult given the many stereotypes of failures of the program in the past, but the stakes are too high for the EPA not to make Herculean efforts to gain greater congressional trust. Shortly after the beginning of the Bush Administration, the Agency began a very serious effort in this regard, and such sensitivity to congressional concerns should continue.

With regard to the financial liability of those companies which originally generated the abandoned wastes years and even decades ago, the Congress has decreed that the companies shall now pay the bill for cleanups. Is this really fair? Should a company which violated no law when it disposed of wastes in the 1950s now be required to go back and make amends? Perhaps after enactment of the National Environmental Policy Act of 1969, all companies should have known better than simply to dump their chemicals on the doorsteps of others. But prior to that environmental awakening, should they have foreseen the urban sprawl that would soon surround those previously desolate fields which no one used to care about? Should they have been able to predict that science would uncover new insights about the toxicity of small traces of chemicals and about the slow but steady movements of chemicals below the Earth's surface? The Congress has answered "yes" to these questions, and it is unlikely that it will change its views. Thus, the EPA and the Department of Justice have the practical problem of retracing history to find the responsible parties and then forcing them to assume their financial liabilities.

As to emergency situations, the only approach that makes sense is "shovels now, and lawyers later." For other situations, the EPA, the Congress, and the public need to be patient, and the "enforcement first" approach of the EPA is sound. If years are needed for documenting the cases in order to force responsible parties to pay, then the EPA should take the time.

Superfund is a long-term program, and delays of even several years at some of the sites in sorting out responsible parties will not make a great difference. Companies have an incentive to pay at the time of cleanup, and the EPA should capitalize on this incentive. Specifically, recalcitrant responsible companies can be fined three times the cost of cleanups if the Agency uses the trust fund and then demonstrates the liability of the companies. The current congressional three-year statute of limitations on collecting payments and fines after the cleanup, however, is too short and should be amended to five years since the EPA has only limited personnel to develop the cases to support such collections.

The EPA should rethink how it distributes its efforts among sites that have already been designated as Superfund sites and do not pose an immediate health hazard. If the responsible parties for a site can be easily identified and are prepared to pay cleanup costs, the Agency should not hesitate to move forward without delay in cleaning up that site at minimal cost to the government even though the site may seem to be of relatively low priority. Low-priority sites have to be addressed sooner or later, and there are both political and environmental advantages in reducing the long list of dirty sites as soon as possible. Of course as already mentioned, special attention needs to be given to leaking sites since a few million dollars spent toward the prevention of groundwater contamination can save tens of millions of dollars in cleaning up polluted aquifers.

During the past several years, priorities among sites have been established by the EPA using risk assessments by experts who rank sites around the country based on the likelihood that hazardous discharges could have an impact on people and ecological resources. However, for sites which are not emergency locations, these impacts will not occur for some years, and the traditional approaches to assessing chemical risks are of limited reliability. In addition to the difficulty in predicting the behavior of chemicals in the land, experts simply do not know how to value the ecological resources surrounding the sites. Wastes are blights on the landscape as well as possible long-term threats to human health, and such contaminated areas have many effects on the surrounding physical and human ecology. The experts need to broaden their conception of the meaning of risks to society, risks which extend far beyond near-term, identifiable health and ecological impacts of chemicals.

Twenty years or 100 years from now the demographic patterns of a region surrounding a site will differ significantly depending on whether the site is retained in its current condition, whether the area is modified in some manner to meld the waste with the countryside, or whether the waste is taken away and the land is reclaimed. In most cases there is no way to predict these patterns. Still, one indicator of the negative impact on society of a waste site is the value which owners, residents, and potential purchasers of nearby property place on their property given the presence or absence of the wastes in their current condition.

Thus, a new dimension in prioritizing Superfund sites is suggested. This consideration should complement, and not substitute for, the EPA criterion of "worst" sites first which is based on risk assessments that emphasize the likelihood that toxic pollutants will reach people in the relatively near term. Specifically, at the outset, the federal government should place more of the burden of prioritizing sites within states on the states themselves. Allocation of funds from the trust fund to geographic regions should take into account the severity of the problems in each state, the track record of the state in responding to waste problems, and the economic conditions of the state. In very general terms, the more sites in a state, the higher the federal contribution to cleanups in the state; the greater responsiveness of the state itself in cleaning up sites, the higher the contribution; and the poorer the state, the higher the contribution.

The states in turn should use a widely publicized test as one important criterion in setting their priorities—namely, the importance which the counties and local communities surrounding the sites on the Superfund list themselves attach to cleanups of the sites. The sites located in communities which contribute financial resources to cleanup operations, contributions scaled to the resources available to the community, would receive a higher priority than similar sites located in communities which are not prepared to make financial commitments. The contribution need not be great, but it should be significant. This approach will provide at least some indication of the long-term risks to society, as perceived by the affected people, associated with the sites. This does not mean that health and ecological risks as determined by experts aren't important, for usually they will be the most important criterion for prioritizing sites. However, local residents are often just as expert as the national experts in determining how dirty sites will adversely affect their life-styles and the life-styles of their successors, and this perspective should not be ignored.

Finally, with regard to the level of cleanup, the Congress should mandate a general standard that should apply to all cleanups whether paid for by industry or the trust fund. The EPA should not be given the flexibility to make exceptions. This standard could be based on the best prediction of the status of the site in 100 years, or thereabouts. One hundred years is too short a time horizon, but predicting conditions even that far in advance is fraught with uncertainty. Also, there is the latent hope that technologies might evolve within the next century which could revolutionize the approaches to management of hazardous wastes.

As an example of a standard, the Congress might call for a nondegradation standard—namely, that the background environmental conditions at the fence line of the site will not be degraded due to residual contaminants at the site for at least 100 years. If such a demanding standard requires hauling away the wastes, then they should be hauled away for appropriate disposal by the responsible parties or by the government.

Partial cleanups should not be undertaken, with the exception of erecting barriers to prevent the spread of groundwater contaminants which could complicate delayed cleanups. Once initiated, cleanups should be carried out quickly with sufficient funds set aside in advance so that funding issues will not be a reason for delay. Such an approach may require reducing the number of sites which are in the cleanup phase at any one time because of the personnel constraints on the EPA and the states for supervising cleanups.

I hope that the following dilemma posed by an EPA site manager will be more easily resolved as the Agency begins to give greater attention to the longer-term implications of site cleanups:

> For this site, I have three remedies that should work—that is, should protect human health. One is a conventional containment approach

> and the other two use treatment. One of the treatment methods involves chemical fixation and encapsulation; the other would yield nearly complete destruction of the toxics at the site. Although that remedy provides the most complete treatment and is the most reliable long-term remedy, it is significantly more costly—nearly two times the cost of the other treatment remedy and six times the cost of the containment remedy . . . . I know the lower cost remedy may not be as reliable in the long term, but I don't know how much more to spend on treatment.[6]

The earlier discussion suggests that the most expensive near-term remedy may be the best long-term remedy, even if it means a considerable delay until cleanup begins.

Safe Disposal of Hazardous Wastes

While the Congress designed the Superfund program to clean up wastes which have been discarded in the past without adequate attention to the environmental consequences, the Resource Conservation and Recovery Act establishes a framework for ensuring that wastes will be handled properly in the future.[7]

Hazardous wastes should be controlled from the time they are generated until their final disposal—from cradle to grave. This is how the current procedure works: For wastes which are shipped off the site of a manufacturing plant for disposal, the manufacturer prepares a manifest which is signed by the transporter of the waste. It is again signed when the waste reaches the disposal facility, and then it is returned to the manufacturer. If the company which generated the waste does not receive the manifest back within a specified time, it can take steps to track down the waste and avoid liability for any possible mishandling of the waste.

This law provides the basis for regulating about 320 high-temperature incinerators and an equal number of special landfills which receive hazardous wastes throughout the country, including spoils from Superfund sites. Many incinerators and landfills are managed by the waste generators themselves. They prefer to take care of their own wastes due to a combination of short-term economic calculations and long-term liability concerns. Consequently, only about 20% of the incinerators and 5% of the landfills are commercial operations open to everyone who does not have access to private disposal facilities.

All landfills receiving hazardous wastes must have systems for monitoring nearby groundwater, and the operations must be covered with liability insurance. The newer landfills are encapsulated within two liners, usually involving layers of clay, concrete, and plastics which are designed to withstand chemical leakages. Engineering systems collect liquids which, despite precautions, nevertheless penetrate the liners. In addition, warning systems for detecting leaks during their early stages are required.

Of course, as we will discuss in a later chapter, the best solution to hazardous waste disposal problems is not to generate the wastes in the first place—a "low-waste" manufacturing strategy. Also, recycling and reuse are obvious ways to cut down on wastes for disposal. These alternatives should be pursued whenever practicable.

If wastes must be taken from a chemical plant, a steel mill, a print shop, or any other facility for disposal, treatment technologies can often reduce the waste volume or render the wastes harmless. For example, precipitation is a technique that removes dissolved chemicals from liquids. Neutralization reduces the acidity or alkalinity of wastes to produce more neutral conditions. Ion exchange is used to remove organic ions from a solution. Oxidation/reduction breaks chemical bonds to detoxify chemicals such as cyanide wastes. Physical treatment can segregate harmful elements from less worrisome chemicals. Incineration destroys wastes at high temperatures. Solidification reduces the migratory potential of waste constituents.

Recent regulations now prohibit the burial of many chemicals in the ground. They are considered too toxic and too persistent for this type of permanent disposal. Many other wastes must be pretreated as described above to reduce their toxicity before the refuse can be placed in a hazardous waste landfill. Frequently, incineration which effectively destroys toxic organic components is selected as the pretreatment method although the other methods are also regularly used. There is now a total ban on placing bulk liquids in waste sites unless they are packed in containers with absorbents to reduce their migratory potential. Another technique to respond to concerns over the escape of liquid wastes is to mix the liquids with fly ash which acts as an absorbent and then to handle the resulting mixture as a solid material.

Despite efforts of the government and industry to reduce the amounts of hazardous wastes and to impose increasing strictures on land disposal of wastes, for the foreseeable future land burial will be the only economically feasible option for substantial amounts of toxic wastes. In this regard, many liquid wastes are placed in surface impoundments where the waste volumes shrink as the water content evaporates. This technique is frequently an important intermediate step toward permanent disposal.

Much of the liquid waste generated nationally is associated with oil extraction and processing in Texas and the South. It is injected into wells one mile or more deep into the earth as a means of permanent disposal. To date there is no evidence that these liquids migrate and pose subsurface problems. However, many environmentalists are concerned over eventual migration of the wastes into both deep and shallow aquifers in, say, 100 years. They fervently believe that this method of deep disposal should cease although they have no economically feasible alternative to offer at present.

The law also establishes general ground rules for the operation of the tens of thousands of municipal landfills that dot every community throughout the country. Unfortunately, over the years many communities have deposited mixes of household trash and toxic waste in dumps that are little more than large holes in the ground. Even today segregation among the individual components in waste streams collected by municipalities—employing a pseudoscience called garbology—is too often the exception rather than the rule.

Municipal waste is diverse. It contains some materials that can be recycled and others that cannot, some that burn and some that do not, and some that should be buried and some that should not. On a national basis, more than 40% of municipal solid waste is paper and paperboard; 18% yard wastes; 8% glass; 8% metals; 6% plastics; 8% rubber, leather, textiles, and wood; 8% food wastes; and 2% miscellaneous wastes.[8]

Leaking municipal landfills are widespread, and many have been declared Superfund sites. However, the problems have not been capped. Every day wastes containing toxic chemicals, sometimes in liquid form, continue to pile up in many landfills which were never designed to contain such wastes. This co-disposal of municipal and hazardous waste is generally prohibited, but the practices of refuse collection and disposal still lag regulatory requirements.

More than one-third of the nation's municipal landfills will be full within the next decade. While the EPA believes that less municipal waste will be placed in landfills as recycling regains its popularity of many years ago, the Agency still predicts that very large quantities of chemical wastes will be deposited in municipal landfills even after the turn of the century. The need to upgrade the requirements for containment capabilities of future local landfills seems clear, and every community will have to shoulder much of the financial burden for new disposal sites. However, this burden for future landfills seems small compared to the more worrisome problem of uncontrolled leaking of toxic chemicals in the decades ahead from many of the poorly designed dump sites of the past.

One of the most difficult environmental issues is the siting of landfills, whether they are earmarked for hazardous waste or for municipal waste. Nearby residents want to close operating landfills and unanimously oppose new waste sites. Still, the nation must have large, well-designed sites for receiving huge quantities of segregated hazardous wastes. Almost every community will need new or larger disposal areas which will inevitably receive some toxic materials mixed with the municipal waste.

With regard to commercial hazardous waste facilities, some states with existing facilities are increasingly resistant to importing out-of-state wastes for disposal at these sites. Also, the capacity of some existing commercial landfills is being approached. In the long run, every state will probably need its own sites. Regardless of the proposed locations for new sites, residents of nearby communities will undoubtedly raise objections.

Greater emphasis on the use of public lands owned by the federal or state governments as locations for landfills may help reduce, but not eliminate, opposition to new sites. In every state there are military stations, for example, that have been dormant for many years or are scheduled for closing in the near future. As part of the "peace dividend" from the relaxation in East–West military tensions, some of these facilities which are geologically appropriate should be made available for hazardous waste sites and for municipal landfills. Such locations could be leased to private contractors who would manage the facilities as commercial operations under tight government controls.

Two Hundred Thousand Leaking Tanks

During the past decade Americans have finally realized that the more than one million underground tanks for storing gasoline, heating oil, and other chemicals which punctuate the nation's landscape have finite lifetimes. In describing the problems associated with leaking tanks, an EPA official reported at a conference in 1989: "A gas station explodes in Council Bluffs, Iowa; a shopping center is shut down for more than one week in Durham, North Carolina; more than a thousand people are evacuated in the predawn hours from their homes in Claymont, Delaware; and throughout the country hundreds of drinking water wells are contaminated."

Bringing the problem closer to home, the storage tanks at the gasoline station two blocks from where we lived in Las Vegas began leaking so badly in 1984 that they had to be replaced immediately, disrupting normal shopping patterns in the area. Then, as I landed at the airport in Miami, Florida, a few days later, I encountered a row of trucks pumping aviation gasoline from the ground near the storage tanks for Eastern Airlines. According to the press headlines, gasoline had spilled and leaked around the tanks for many years and finally entered the shallow aquifer near the airport that fed into Miami's drinking water supply.

Sometimes, cleaning up leaks makes economic as well as environmental sense. Several companies in Texas and Oklahoma, for example, have made substantial profits in recovering large quantities of spilled and leaked chemicals around petroleum storage tanks in the region.

The unfortunate incidents cited by the EPA and others are not the results of careless disposal of chemicals. Rather, they exemplify the concentration of businesses and farmers on the near-term future and not on problems that are not easily foreseen. Most of us have believed that steel tanks are surely adequate to store gasoline or heating oil; and if there were an underground leak, the liquid wouldn't go very far. How wrong we have been. The potential costs to the nation of cleaning up are very high.

In 1984, the Congress responded to the many reports of leaking storage tanks. New legislation established a program to improve standards for owning and operating tanks, to help detect leaks as soon as possible, and to provide the legal and financial tools for prompt cleanup of escaped liquids. When corrective actions seem necessary, the EPA can require owners and operators to test their tanks for leaks, to excavate sites and assess the extent of contamination, and to clean up the contaminated soil and groundwater.

However, many leaking tanks are discovered at abandoned sites, and the responsible parties have disappeared long ago. In other cases, the owners or operators may not be able to afford the cleanups, or they may refuse to take action. Thus, another fund was created to enable the EPA or the states to take immediate action when necessary to clean up the problems. The fund relies on a small federal tax on certain petroleum products, primarily motor fuels. However, the fund is not a bailout, and owners and operators remain liable for the costs which will probably become higher if the government rather than the responsible parties undertakes the remedial actions.

Owners and operators of underground storage tanks are now required to maintain the financial capability to clean up leaks through liability insurance or other means. For petroleum production, refining, and marketing facilities, for example, the Congress has established minimum coverage levels at $1 million for each occurrence of a leak.

How can the EPA, the states, and the nation cope with the huge number of tanks which have already been discovered and will be discovered? The early estimate of 200,000 leaking tanks may only be the beginning. Many of the tanks are owned by individuals who have neither the technical wherewithal nor the financial resources, and in some cases not even the personal commitment, to adequately monitor the state of their tanks for the indefinite future.

In order to increase the incentives for greater diligence toward maintaining the integrity of underground tanks, some states only permit transfers of a title for commercial property if the state environmental agency has certified that the property is free of leaking tanks. Other states should be encouraged to adopt similar programs which place significant economic value on protected groundwater resources. This type of requirement might be extended to include agricultural and residential property as well.

The Fears and the Reality of Nuclear Waste

Some citizens are frightened of anything nuclear. Not surprisingly, nuclear waste disposal is one of the most politically volatile environmental issues facing the nation.

Since the accident at Three Mile Island, nuclear power and nuclear weapons have become inextricably linked in the minds of rabid antinuclear activists. They have forged alliances with many other calmer but still concerned Americans who believe that both of these products of World War II pose serious threats to our survival. Few federal, state, or local politicians can afford to ignore vocal constituencies who vehemently oppose bringing the waste by-products of weapons or reactors into their jurisdictions. They perceive no economic value from such waste disposal industries, only lots of headaches.

However, if nuclear power is to become a more significant component in the energy mix of this country, or indeed retain its current place as an important contributor to our electrical energy, the issues surrounding the disposal of nuclear fuel rods which have been used up in power reactors and are impregnated with radioactive contaminants—usually called high-level waste—must be resolved. Even if nuclear power is abandoned, the nation will have to cope with the current inventory of used fuel rods for centuries to come. The temporary solution since the late 1950s has been to simply retain the high-level waste at each of the 110 reactors around the country. This shortsighted approach, which has been the only politically feasible solution to date, is expensive and will require construction of additional storage capacity at some locations as sites reach their storage limits. Meanwhile, as nuclear wastes accumulate, local political anxieties heighten at many of the reactor locations.

Other types of nuclear waste generated from different sources are also important, and problems encountered with these wastes often confuse the debate about nuclear power. In particular, many hospitals and commercial facilities use radioactive isotopes for diagnosing the conditions of both people and materials. Scientific laboratories use radioactive tracers and nuclear irradiation techniques for understanding the laws of physics and chemistry. Luminescent dials and signs rely on the radioactive properties of some materials. Occasionally, wastes from such industrial, medical, or research activities are not handled properly, and low but detectable levels of radioactive wastes turn up in junkyards or dumps not designed to handle such materials. Over the years radioactive contaminants have crept into many Superfund sites, thus complicating cleanup procedures. These concerns must be addressed but should be kept separate from the debate over nuclear power.

Three key issues with both economic and environmental dimensions will largely determine the future viability of nuclear power. They are demonstrated safe performance of reactors as evidenced by prevention of accidents which might contaminate workers or nearby residents, the safe decommissioning of nuclear stations after their useful lifetimes of 40 or 50 years, and as noted, environmentally sound disposal of fuel rods which have accumulated high levels of radioactive contaminants.

Prevention of reactor meltdowns and other types of accidents is discussed in a later chapter. Particular emphasis is placed both on the importance of designing and testing reactors which will shut down automatically in the event of human or mechanical failures and on the necessity to improve the capabilities of operating personnel to respond to unexpected events, even with reactors believed to be completely safe.

As far as the decommissioning of old power reactors is concerned, the initial American experience is encouraging. An early power reactor which operated at Shippingsport near Pittsburgh, Pennsylvania, has been successfully dismantled. Building on this initial experience, the industry will undoubtedly begin decommissioning other reactors during the next decade, confident that the technical problems will be easily resolved. Of course some elements of the public will always object to the selection of any location as the final resting place for old reactor vessels, pipes, and other large contaminated components that are removed from nuclear reactor sites. Still, outmoded reactor complexes can be dismantled, decontaminated, and then used for other industrial purposes provided they have not been the scenes of major accidents such as Chernobyl which has become a permanent nuclear graveyard—a testimonial to a unique Soviet reactor design that paid little heed to safety requirements.

At the present time, disposal of depleted nuclear fuel rods is a principal nexus for joining the public debate over the future of nuclear power. Las Vegas is in the center of the debate. I, like most Nevadans, have been exposed to hundreds of newspaper accounts and many hours of television broadcasts about high-level nuclear waste disposal.

Tempers run high and engineering concepts are constantly challenged in southern Nevada when the conversation turns to the disposal of high-level wastes. Since the early 1980s when the federal government began to settle on Yucca Mountain to the north of Las Vegas as the permanent cemetery for spent fuel rods, my days on the tennis courts of several casinos and my evenings at social gatherings have been frequently punctuated with expert citizen advice to all who will listen on how to change the policies of the Department of Energy on radioactive waste disposal. This advice is usually very simple: Take the waste somewhere else. I disagree with this advice.

To provide a perspective, the technical problems associated with disposal of nuclear fuel rods are minor in comparison with the problems of chemical wastes. The volume of these nuclear wastes is relatively small, and even if the nation increases its dependence on nuclear power the growth will remain small in comparison with the huge volumes of chemical wastes already in the ground and being generated each year. In contrast to its uncertain assessments of chemical pollution, our government knows exactly where the fuel rods are located. They are concentrated at a relatively small number of locations, and there are no undiscovered burial sites. Our specialists know how to monitor for the presence of radioactivity. They do not need to launch a major research program to develop new lines of devices for detecting and measuring radiation.

Having studied for 40 years the health impacts of radiation on the Japanese population following the nuclear detonations in Hiroshima and Nagasaki, American and Japanese doctors know the degree of danger associated with human exposure to radiation. They need not rely on highly uncertain extrapolations to humans from the reactions of laboratory rats and mice as the basis for their medical judgments.

Yucca Mountain is a barren patch of desert 100 miles northwest of Las Vegas where even the jackrabbits have difficulty finding companions. For many years it has been off-limits to wandering prospectors, to lost campers, and now to antinuclear demonstrators since it is at the edge of the Nevada test site, a high-security area where nuclear weapons are tested. Rainfall is a rarity, and many hundreds of feet below seemingly impenetrable volcanic rock groundwater flows with a speed so slow that it is difficult to measure. The mountain rises perhaps 1000 feet above the desert floor with a special appeal for desert artists who are taken with picturesque sunsets.

Sometimes the discussion at Las Vegas cocktail parties turns from showgirls competing in bicycle races, from heavyweight title fights, and from Wayne Newton to the dangers of transporting nuclear fuel rods along the highways of Nevada. In reality, the risks in transporting a limited number of fuel rods in specially designed and repeatedly tested containers are minimal. The lead canisters simply will not split open regardless of impact.

Meanwhile, every day the residents of Las Vegas live with the risks of transportation accidents involving chemicals being carried through the city in trucks of all descriptions. However, to appease Nevadans, the federal government has proposed to build a special rail line across federal lands for transporting nuclear wastes from the border of the state to Yucca Mountain.

The weak link in the case of the Department of Energy for placing high-level wastes in Yucca Mountain is the burial method. The idea is to permanently emplant the fuel rods in deep shafts where the radioactivity can decay over many centuries in a manner that will not affect the environment. They are to be sealed forever and eventually become an integral part of the earth's mass. This concept of permanent geological burial was developed 30 years ago. Geoscientists have spent hundreds of millions of dollars trying to persuade political leaders that nuclear wastes, buried in appropriate locations, will not bother anyone for 10,000 years.[9]

At Yucca Mountain, the proposed burial site is sufficiently high above the groundwater that even if leakage begins, the time for migration of the radioactive liquid through the volcanic rocks to the aquifer will be thousands of years, argue the government experts. Also, they contend that the likelihood of an earthquake disrupting the repository is so remote as to be negligible.

Interminable arguments are now under way between these experts of the federal government and other experts mobilized by the state who challenge the underlying concepts that led to the choice of Yucca Mountain. The state argues: Can you really be sure that there will not be an earthquake? Even if the earthquake doesn't impact on the repository directly, couldn't it change groundwater patterns and increase the vulnerability of the water to leaks? Couldn't population growth in areas near the site and attendant withdrawals of groundwater for drinking and agriculture change groundwater flow patterns? Couldn't miscalculations result in burial practices that generate excessive amounts of heat which would lead to dangers of combustion in the repository?

Trying to predict conditions hundreds of years into the future, and in this case thousands of years, is plagued with uncertainties. There is always a chance, albeit very small, that an earthquake could occur in this region which has been historically quite free of earthquakes. Furthermore, the costs of preparing the repository, emplacing wastes in the repository, and maintaining surveillance of the conditions hundreds of feet under the ground would take billions of dollars within the first few years.

An approach that would seem more acceptable technically and politically and that would be cheaper in both the short term and the long term is simply to store the high-level waste in lead containers on the surface of the desert. They could be appropriately spaced and cooled by the air with no chance of mechanical failures and no danger of building up excessive heat. An earthquake could of course disrupt the site and perhaps scatter the canisters over the countryside. However, the site would be relatively easy to restore. The canisters would probably remain intact, or at worst the contamination from those that cracked under the great pressure would be localized. But compare these consequences to the disruption of underground caverns with highly uncertain subsurface consequences which could not be put back in order. Further, sufficient distances between the wastes and the fence line could ensure that the radiation levels off the site would not even be measurable. There certainly is no shortage of space near Yucca Mountain, and security will be extensive for the indefinite future given the military secrets buried in the adjacent weapons testing area. Of course, some nuclear-phobic members of the public might find burial more comfortable than having easily photographed exposed waste—regardless of the technical considerations. But nothing short of impossible transformation of nuclear waste into harmless dust will ever satisfy rabid antinuclear forces.

In the past, the Department of Energy has considered such an approach, although not in Nevada. The department has referred to collecting high-level waste from around the country and then placing it in surface storage at a single location, preferably near Oak Ridge, Tennessee, as an "interim" retrievable disposal method for perhaps 20 years pending permanent geological burial. Unfortunately, the department's two-step process of interim and permanent disposal in different locations doubles the number of politicians who oppose the scheme.[10]

After spending the past several decades trying to develop an acceptable approach to permanent geological burial, the department should abandon such a concept, at least for the time being. Interim "retrievable" storage aboveground should be considered the goal for the next century. If technologies of the future offer new opportunities either for alternative disposal approaches or for unanticipated future use of the materials embedded in the fuel rods, the wastes would be readily accessible. The Nevada desert has everything required for such storage.

Two common arguments against surface storage have been (1) vandals or terrorists could disturb the wastes and (2) nuclear devices on incoming missiles could hit the site and scatter radioactive debris. However, vandalism in the Yucca Mountain region is not easy, given the remoteness of the area and the security procedures nearby at the Nevada test site. Meanwhile, a crowded Caesar's Palace on the Las Vegas strip offers a far easier and more lucrative target for terrorists than a high-security area in the desert. As to a surface-storage site being a sitting duck for a nuclear attack, I would rather have the incoming nuclear weapon hit a site 100 miles north of the city and take my chances that the lead caskets will contain most of the stored waste than having ground zero be downtown Las Vegas. The suggested threat scenarios depict extraordinarily inefficient ways to cause harm through the spreading of nuclear debris.

Perhaps the most popular argument offered by local politicians against depositing waste in the state is that Nevada has done more than its share for the nuclear effort. Indeed, more than 700 underground caverns filled with radioactive debris from underground tests punctuate the Nevada test site. Each one of these hot sites more than qualifies as a Superfund site, and the tests are continuing. Not surprisingly, with at least 5000 jobs directly dependent on continued testing—let alone associated service jobs in Las Vegas—the same politicians who oppose a nuclear repository in Nevada support increasing the state's nuclear burden through continued testing which creates the most undesirable type of debris. Similar economic incentives will be needed together with Washington political muscle, to begin to mollify the Nevada opponents of a waste repository in their state.

The political realities were crisply summarized by a columnist in the Las Vegas *Review Journal* in March 1990 as follows:

> . . . the state's top officials are so adamantly anti-dump that they fight it in court and use guerilla bureaucracy to delay study of the site. . . . The federal government and the nuclear lobby had it made in 1974, but they screwed up their chance and now think they can force the dump on Nevada. It's a big miscalculation. Nevadans can stand nuclear testing and, at least at one time, could tolerate the idea of nuclear waste storage. But not now, not ever, will Nevadans take kindly to being bullied by the federal government. That's something that makes Nevadans downright unreasonable.[11]

Another viewpoint by a Las Vegas resident was published several days later:

> The Nevada Test Site is already a repository of nuclear waste, and the addition of new material from around the country seems of small consequence. After at least 700 acknowledged explosions above and below ground, it would appear to any reasonable thinking individual that contamination of the site exists. The trained personnel already in place plus the equipment, housing, facilities, guards, transportation, and knowledgeable companies that have done this testing for years seem an excellent investment to continue to monitor and oversee this storage of nuclear material. As pressure continues to mount for the cessation of all nuclear testing, it seems a waste to allow this enormous expenditure of taxpayer dollars, manpower, and expert knowledge not to be put to use.[12]

From the Washington vantage point, southern Nevada is the obvious choice for the repository. No other location can combine physical isolation, a large workforce highly experienced in handling nuclear materials, and readily available security services. Historically, Nevadans have accepted environmental contamination of a remote desert area through nuclear testing, and the added burden of placing nuclear wastes in a nearby location is small.

In the end, the political forces in Nevada will probably be overwhelmed. The problems of nuclear waste are simply too important and will not go away. However, the Nevada politicians will undoubtedly extract from Washington substantial financial benefit for the state in exchange for allowing this expansion of nuclear activities in the desert.

Environmental Neglect at Nuclear Weapons Plants

For 40 years officials of the U.S. government responsible for the nuclear weapons program were protected by the shield of national security in presenting their case to the nation that nuclear weapons could be produced quickly and *safely* and at a reasonable cost. While every president since Harry Truman has taken an intense interest in the capabilities of nuclear weapons to destroy the Soviet Union's environment, prior to the ascendancy of George Bush not a single one bothered to investigate what nuclear weapon production was doing to our environment. Had one of our presidents looked inward as well as outward, he would have seen hundreds of examples of environmental abuse—large areas of soil laced with plutonium which can be resuspended in the air, radioactive liquids leaking from storage tanks into underground aquifers, and contaminated vehicles and other equipment which were simply abandoned and covered with loose dirt. Such abuses were too often reflected in an attitude of, "Dump the wastes out back, and we'll worry about them later."

Largely as the result of citizen pressure, "glasnost" has come to the nuclear weapons complex. In a rapid turnaround, many congressional leaders as well as senior officials of the executive branch who ignored the problems for many years have now become the nation's most vocal environmentalists. "Environmental protection first, weapons production second," proclaims the Secretary of Energy while relying on advice from many experts who have been quiet on the issue for decades. Meanwhile, the Department of Defense is having an increasingly difficult time making the case that "The Russians are coming, and we need to expand our nuclear stockpile of 22,000 weapons."

During my time as director of the EPA's environmental advisory services on nuclear testing at the Nevada test site from 1980 to 1985, I was often disturbed by the views of some of the managers of our nation's weapons production, and particularly Washington-based officials who visited the desert. Their overriding objective was to develop more efficient weapons, and they did not appreciate having impediments put in their way, including environmental assessments which might complicate their task. Some thrived on their image as nuclear cowboys, symbolized in the readily available government pickup trucks, subsidized steak dinners, and large budgets to support projects with questionable justifications.

As noted earlier, the Nevada test site was designed to be an environmental wasteland. The possibility of preserving the area for any purpose other than nuclear waste activities disappeared with the first nuclear tests in the 1950s. The EPA's major preoccupation, therefore, has been to help ensure that radioactivity does not leave the site. While not enthusiastic about the EPA's intrusion into their affairs, the weapons managers have recognized the public relations value of having the EPA on their side; and they try very hard to ensure that the EPA can endorse their approaches to off-site safety. As history is now revealing, the record of environmental consciousness at the Nevada test site, aside from the deliberate contamination of the subsurface environment every time a weapon is detonated, has been very high in comparison to the records at the other dozen or so sites of the nuclear weapons complex. Still, the mind-set of "better weapons whatever the price" has permeated the Las Vegas area as well as the other sites for many years.

In the early 1980s I traveled to Idaho to review environmental programs at another large test facility of the Department of Energy. My suspicions about the incompatibility of weapons activities as they were then conducted and environmental protection were confirmed. A problem of particular concern was the leakage of radioactive tritium into a large underlying aquifer which feeds into the Snake River. The tritium had already migrated several miles off the federal property, and the nearby communities had become upset that the pollutant would contaminate their drinking water supplies.

Our host at the manufacturing facility responsible for the tritium leakage told us not to worry about the stories of off-site environmental problems. The environmental groups in the region had carefully reviewed the situation and had concluded that the tritium leakage was insignificant. When I expressed interest in the involvement of environmental groups, he smiled. He then proudly stated that he was an active member of the local Audubon Society and had played an important role in the "independent" environmental critique of the manufacturing facility where he was employed.

In 1989 the task of cleaning up the refuse from weapons activities at Rocky Flats in Colorado, Hanford in Washington, Fernald in Ohio, and the other nuclear weapons facilities (occupying a territory larger than the states of Delaware and Rhode Island combined) began in earnest. Projections are that the price tag over the next 20 years will be about $250 billion of federal funds. This cost boggles the minds of environmentalists who often are satisfied with grants of $10,000, $100,000, and occasionally $10 million to restore a polluted area. However, the members of the military–industrial complex, accustomed to annual defense budgets which significantly exceed $250 billion, took the cleanup costs in stride.[13]

John Glenn, chairman of the Senate Subcommittee on Governmental Affairs and a leading activist in promoting environmental awareness throughout the military weapons complex, offers the following guidelines: First, the Department of Energy in developing and producing weapons should operate on a pay-as-you-go basis, including the costs of health, safety, and environmental protection. More realistic assessments of the need for additional nuclear materials and alternatives to increases in our weapons arsenal should be explored. Greater oversight by competent scientists and representatives of the public of nuclear weapons activities is needed both in the management of facilities and in the design of radiation research programs. Finally, steps must be taken to ensure that once production activities at a facility end, there will not be residual environmental problems.[14]

The following admonition of Senator Glenn is right on target: ". . . the notion of harming the health and safety of large numbers of Americans in order to produce weapons makes a mockery of the phrase 'national security.' We must face up to this reality by bringing America's nuclear weapons industry into the modern era and making it accountable to the citizens it is designed to protect."[14]

Since the advent of nuclear weapons, the Department of Defense has failed to consider adequately the environmental costs of its military activities. Now the nation's defense budget, and not its environmental budget, should be charged the costs of cleaning up the debris from the nuclear weapons complex.

Protecting Our Groundwater Resources

Accompanying the dramatic increase since the early 1980s in public anxiety over radioactive and hazardous waste disposal practices has been a growing awareness within and outside government agencies that the quality of America's groundwater is slowly but steadily deteriorating. A key concern at almost every waste disposal site is protection of the groundwater under and near the site. Unfortunately, the documented cases of leaking wastes which have contaminated underground drinking water supplies are on the increase in many regions.

People depend on groundwater in every state. It may be only a few feet below the surface or it may be hundreds of feet into the Earth's crust. It currently provides one-fourth of the water used in the country. One-half of the American population including 97% of the residents of rural areas obtain their drinking water from underground aquifers. Groundwater provides 40% percent of agricultural irrigation water and a considerable portion of water used by industry. Also, it nourishes aquatic ecosystems which are valued for their fish, wildlife, and recreation opportunities. In periods of drought, groundwater is particularly important in ensuring a continuing supply of fresh water for many lakes, rivers, wetlands, and estuaries.

Most groundwater in the United States is clean and available in adequate quantities to meet our needs. The nation as a whole is clearly not facing a groundwater crisis. However, in a few regions, the withdrawals of groundwater exceed replenishments. Of our immediate interest, a wide array of contaminants have been detected in many areas. Agricultural fertilizers, pesticides, heavy metals, and solvents have received the most publicity as groundwater contaminants. Meanwhile, government surveys have found many more chemicals, totaling over 200, in the nation's groundwater.

In most contaminated subsurface areas, the experts have discovered only minute levels of these substances. Further, most cases of serious groundwater pollution are highly localized with contamination plumes seldom being more than one or two miles in length. The plumes usually can be traced to chemicals escaping from wastes sites, spills of chemicals, leaking underground chemical tanks, old septic tanks, or excessive use of agricultural chemicals.

Unfortunately, many of the patches of groundwater which are

contaminated with man-made chemicals are located in densely populated areas where groundwater is an important source of drinking water. In some localities, officials have closed contaminated wells. Such closures already have affected millions of consumers. For the foreseeable future, this pattern of localized pollution of groundwater will probably continue to intensify, and health inspectors will continue to close wells. Even if additional chemicals were not deposited on the land in an uncontrolled manner, some areas of soils are already saturated with chemicals which will eventually reach the water table. The costs of halting the movement of contaminants toward underground resources, pumping out contaminated groundwater, cleansing it at the surface of the ground, and reinjecting it back into an aquifer are very high and often prohibitive; and we must accept the inevitability of additional groundwater contamination.

As to underground contamination that can cover large areas, the problems of groundwater pollution from farming practices are finally being recognized throughout the country. Nitrate fertilizers and soil additives such as gypsum and sulfur are found in groundwater in some agricultural areas. High levels of salinity are induced through reuse of irrigation waters that collect and concentrate chlorides. These chlorides may occur naturally or may be constituents of agricultural chemicals. Animal wastes often contribute bacteria and salts to groundwater. Of course, excessive use of pesticides remains at the top of the list of concerns, and this problem is discussed in the next chapter.

Other sources of groundwater contamination that can affect large areas include acid drainage from mining areas, runoff from highways of deicing salts, and seepage of wastes from leaking septic tanks. Also, in some areas groundwater is so close to the Earth's surface that it intermingles freely with surface waters receiving all of the common runoff pollutants that plague streams and rivers.

At the national level, several types of actions are designed to protect groundwater resources. First, we have already discussed efforts to contain chemical wastes and chemical storage facilities and to clean up those sites where chemicals leak into groundwater. Second, federal and state agencies restrict the use of pesticides and other toxic chemicals which are placed on the land and do not degrade for many years, with the restrictions designed to ensure that they will not reach groundwater. An additional program calls for a few large aquifers, often spreading across large portions of several states, to be designated as particularly valuable resources of drinking water. Federally financed projects which might impact on the aquifers, such as construction and water development schemes, can only be undertaken after studies confirm that these activities will not adversely affect the quality of the aquifers. Cutting across all of these activities are federal standards for drinking water—namely, the levels of chemical contaminants which pose no threat to health. These levels are generally used as a guide as to when groundwater is clean enough.[15]

Of comparable importance are the actions taken by the states to protect groundwater. Many agricultural states, such as Nebraska, are adopting strong stands to limit the excessive use of agricultural chemicals and to discourage farming practices which permit runoff water to drain into groundwater. In a few areas such as Long Island, geographical zones are designated according to the present condition of the groundwater—such as pristine, partially contaminated but usable, and contaminated. Activities permitted in each of the zones are constrained by regulations to prevent further degradation. For example, waste sites may be located only in zones where the groundwater is already contaminated, and even there they must be carefully monitored. In the pristine zones, activities are sharply limited to those with little possibility of spilling chemicals into the subsurface environment.

The protection of groundwater is first and foremost a land-use issue. Historically, all levels of government have been hesitant to tell citizens how they can use their land. In recent decades, the responsibility for local zoning restrictions has been aggressively pursued in communities throughout the country. While activities funded by federal and state agencies do influence these local deliberations, the agencies have usually stood aside during the detailed planning of areas which are primarily private property.

Groundwater resources typically extend far beyond the boundaries of individual communities. Few communities have the technical wherewithal to assess the likelihood of threats to groundwater in their immediate vicinity and the feasibility of abating these dangers, let alone the problems of movement of underground water into other areas. Furthermore, communities usually have short time horizons of, say, 50 years whereas groundwater pollution induced today will be with us for centuries into the future. The states, with a broader perspective, can make important contributions to local decisions.

About 20 years ago, the federal government awakened to the accelerating pace of real estate development which threatened the ecology of the nation's coastal areas. This was also a land-use issue involving private property, and the decisions of one coastal community affected others as well. Thus, the special role that the states could play was recognized. The federal government has provided grants to states for planning coastal development in accordance with very loosely articulated nationwide objectives. Also, tied to these planning activities are a variety of federally funded programs which impact on coastal development. Thus, many states have adopted a protective stance to the regulation of private property. Lessons from this experience seem relevant to the approach for groundwater protection.

Indeed, every state has a designated groundwater office. The financial support and the technical advice of the federal government are very helpful to these offices. Also, a firm commitment of the many interested federal agencies to try to shape the local programs which they fund within the overall regulatory frameworks developed by these offices can provide greater rationality in both the national and local approaches to groundwater protection.

However, even with unprecedented success of federal and state efforts to reverse trends in groundwater contamination during the next few years, the United States will enter the next century with patches of the nation's groundwater polluted beyond hope of recovery. In some localized areas, groundwater supplies must be written off as not recoverable for some uses, and the local populations must become accustomed to this reality.

At the same time, a reasonable goal to establish now is the prevention of further significant degradation after the turn of the century of *any* underground aquifer with a potential for use. Policies and programs during the 1990s should be directed toward minimizing the extent of the contamination that could eventually migrate into aquifers. Strong efforts are needed to keep the number of American citizens who believe they have no choice but to resort to bottled water as low as possible. Currently, more than 20 million of our citizens drink bottled water. Some have rejected tap water in favor of a chic habit. Others prefer the taste of sparkling mountain springs. Still others are convinced that tap water is unhealthy. And some live in areas where they simply have no other choice.

All of the approaches for reducing and controlling wastes discussed in this chapter will impact on the status of the nation's groundwater. Policies for cleaning up improperly discarded wastes, for handling municipal wastes, for improving the integrity of underground storage tanks, and for locating and operating permanent disposal facilities for chemical and nuclear wastes are critical for the preservation of much of the nation's freshwater resources. While other types of environmental threats such as air pollution from factories and cars may be of more immediate concern for human health, in the longer run waste problems must be at the top of the list of environmental protection priorities.

6 The States Begin to Take Charge

The powers not delegated to the United States by the Constitution, nor prohibited by it to the States, are reserved to the States respectively, or to the people.
—Amendment X of the U.S. Constitution

Congress and the Executive Branch have been totally paralyzed. . . . We simply can't afford to wait.
—Environmental Commissioner of the State of New York

State aid: Towns will get rules, but not money, to protect the environment.
—*Cape Cod Times*

A Soft Collar Protects the Chesapeake Bay

"Where are all the scientists?" I asked a colleague from the EPA. We were attending a conference in 1980 at an old Virginia hotel overlooking the mouth of the Chesapeake Bay. The purpose of the gathering was to review the progress of the EPA's $25 million scientific studies of the ecological condition of the Bay. I had recently assumed a new position as the manager of aquatic research within the Agency's headquarters in Washington, and the Chesapeake Bay program was one of my responsibilities. I had expected to see mostly scientists among the 150 participants at the meeting, but we were surrounded primarily by representatives of state, county, and town governments and by environmental activists.

My EPA colleague who had followed the Chesapeake Bay program for several years reassured me that we were at the correct meeting and that the EPA had indeed emphasized science in its studies. He noted that any major field research effort close to Washington, and particularly one targeted on a recreational mecca for Washington politicians, had immediate repercussions that transcended science. Therefore, the domination of the meeting by specialists directly involved in governmental policies at many levels was to be expected, he added. Of greater significance to me, however, were the comments of my colleague that the EPA-sponsored studies would only be successful if the findings were translated into action, and the attendees at the meeting were precisely the types of people who could bring about changes in the amount of pollutants being deposited in the Bay.

The Chesapeake Bay is one of the most bountiful estuaries in the United States. More than 50 rivers interrupt the thousands of miles of shoreline and feed into almost 5000 square miles of water. The fisheries of the Bay provide 95% of the nation's blue crabs and over 50% of its oysters.

Almost 15 million people live in the watershed of the Bay—in large areas of Virginia, Maryland, and Pennsylvania, and in the District of Columbia. The Bay is an important route for international shipping. It is known throughout the world for its sailing and other forms of recreational boating. At the same time, the Bay is the recipient of large amounts of pollutants from cities, from industrial facilities, and from agricultural areas every day.

Returning to the conference, the few scientists in attendance were trying their best to document trends in the environmental quality of the Bay during a 50-year period. They were able to show that during the previous decade fisheries and seagrasses had declined, levels of phosphorous and nitrogen from agricultural fertilizers had risen, and in many areas oxygen had become inadequate for healthy aquatic life. They could pinpoint dramatic increases in metals, pesticides, and other toxic pollutants flowing from urban and rural areas into the Bay. However, the scientists had great difficulty going back several decades with their comparative analyses since scientific data and records were far from complete.

The most persuasive evidence as to the continuing decline in the quality of the Bay came not from the scientists but from a maritime pilot who for 50 years had been guiding large ships over a 100-mile trek from the Atlantic Ocean into the port of Baltimore. A true lover of nature, he presented his observations over the years as to how the wildlife had changed, how clear water had turned gray, and how the Bay had been repeatedly and, in his view, irreversibly scarred from chemical discharges on the land and on the sea. After his lucid and at times emotional presentation, there were no doubters in the audience that the time had long passed for aggressive steps to protect this aquatic treasure.

Protection of the Chesapeake Bay provides an excellent example of the central role which must be assumed by state governments in correcting environmental problems. The federal government can provide funding to support local efforts. The agencies in Washington also can support research that will help clarify the seriousness of the problems, and they can place limitations on certain types of polluting activities. But corrective actions of the magnitude required to preserve the Bay can only come about if the states and the local jurisdictions fully commit themselves to environmental protection. In this case, more than 100 local jurisdictions ring the Bay and its principal tributaries, and only the state governments are in a position to galvanize local actions so that all pull in the same direction.

Turning specifically to the state of Maryland, many efforts are currently under way to protect the Bay from pollution in general and from chemical contaminants in particular. The state has steadily upgraded municipal sewage treatment plants and has vigorously enforced regulations limiting chemical discharges from industrial plants into rivers and streams. It requires industry to cleanse its water before discharging it into municipal wastewater systems. Also, the state limits rural runoff into the Bay through erosion control programs. These are all standard approaches used to varying degrees to control pollution throughout the country.

An important and novel centerpiece of Maryland's efforts to protect the Bay is the direct regulation of a 1000-foot strip of land around its portion of the Bay and adjacent tidal marshes, an approach that has also been adopted by Virginia. All counties and municipalities within this protective collar must have zoning and development programs, approved by a state commission, which minimize the adverse effects of growth. Wildlife habitats, soil, endangered species, tidal and nontidal wetlands, forests, and fish-spawning areas are of special concern. This bold regulation by the state, in cooperation with local authorities, of the use of private land is both unusual and effective. While some property owners have been apprehensive of governmental encroachment on their activities, most have accepted the approach as improving the condition of the Bay and thereby raising the value of their property in the long run.[1]

Of course, many bays and lakes throughout the country are not ringed by soft environmental collars. However, every state has its own initiatives to protect environmental resources—rules on the spraying of chemicals, guidelines to improve the management of farming areas, computer mapping programs to identify environmentally sensitive areas, and many other approaches. Sometimes the federal government encourages and supports such initiatives, and sometimes officials in Washington learn about them only after they have been adopted. The remainder of this chapter discusses some of these approaches and describes how the states are increasingly taking the lead in protecting their resources. More often than not, state leaders feel that Washington is either out of touch with their real problems or is too bogged down in administrative procedures to respond to local needs in a timely manner. Central to the discussions are the issues of funding and control of programs as environmental problems test the concept of "states' rights" as never before.

Setting the Environmental Agenda and Paying the Bill

Looking more broadly at local environmental issues, in 1989 *Newsweek* underscored the differences in approaches to environmental protection being pursued throughout the country—from the Chesapeake to the San Francisco Bay, from the Black Hills to the Everglades, and from the Sierra Nevadas to the Appalachian Mountains:

> Are you sick of inhaling the gasoline fumes wafting from service stations? You'll breathe easier in the northeast where eight states have decreed that only a new "clean" gasoline can be sold. Fed up with utility plants that spew out the ingredients of acid rain? Move to Wisconsin or Minnesota, which strictly limit such emissions. Worried about radiation leaking from a nuclear power station? You'll sleep easier in Illinois where the nuclear monitoring program surpasses any federal effort.[2]

The magazine trumpets the dawning of a new age of environmental federalism, proclaiming that the states have passed more stringent controls on pollution than the Congress ever considered. For example, *every* state now regulates emissions of toxic chemicals into the atmosphere. *Newsweek* concludes that the states are forging ahead on their own because Washington simply will not aggressively champion a number of environmental measures even on issues which have causes and consequences far beyond any state's borders. While this criticism of federal inaction seems excessively harsh, the perception of environmental footdragging in Washington, and particularly with regard to toxic chemicals, is certainly widespread.[2]

At the heart of the never-ending debates throughout the country as to the division of responsibility between federal and state environmental agencies is the question of who will provide the funds for state environmental activities. In general, the state agencies would like to have full authority to control activities within their boundaries. However, they would also like to have all of the funds for their programs come from Washington. They recognize that they can have neither, and a tug of war proceeds on many fronts.

Since the earliest days of our nation, the federal government has assumed responsibility for issues which impinge directly on interstate commerce. Now the definition of interstate commerce is becoming blurred as environmental policies sooner or later affect the prices of most goods which are produced in one state and marketed in others. Also, pollutants move in often invisible ways from one state to another. The uncertainty dividing interstate and intrastate environmental actions clouds both the regulatory and the funding responsibilities for many state environmental initiatives.

In addition, state and local politicians are sometimes uncomfortable dealing with emotionally charged environmental issues which are extraordinarily complicated in their origins and in their solutions. Frequently they would like to pass these environmental hot potatoes to Washington for resolution, and they seldom advocate the exclusion of federal involvement in any of their environmental programs.

Finally, every state recognizes the importance of research in clarifying the dimensions of environmental problems. They all want access to the huge federal research efforts, and they want scientific efforts oriented toward their specific problems.

Returning to the issue of financing, the National Governors Association has pointed out that during the next decade the costs of environmental programs administered by federal, state, and local agencies will grow by more than $20 billion, with the bulk of the financial burden coming to rest in the state capitals and county and local seats. According to the governors, their financial commitments for environmental protection are already growing rapidly. Most states are struggling to meet the costs of many existing environmental initiatives, they emphasize. Also, the governors note that the states are receiving a declining share of the EPA's budget. While many states have increased their appropriations from general revenues, these resources are not adequate to cover program costs, they add.[3]

Exemplifying the bargaining positions adopted by the states in their quest for larger federal funding with less federal control is the following summary of the declaration of the National Conference of State Legislatures in 1989 concerning activities under the Clean Water Act:

1. The Congress and the Bush Administration should provide *all* of the funds authorized in the 1987 Clean Water Act reauthorization to treat wastewater and to control pollution from runoff.
2. States must be given increased flexibility in determining the most beneficial and cost-effective use of federal funds provided for wastewater treatment.
3. The states should be given more flexibility in the use of federal funds to support their programs for cleaning up pollution spills and other emergency situations, including removal of federal limitations on the percentages of funds that can be used for administration of such programs.[4]

The funding alternatives to federal grants and general revenues which are used most widely by the states are special fees, taxes, bonds, and revolving loan funds earmarked for environmental programs. Other fund-raising approaches include fines and penalties for violations of environmental regulations, private contributions for environmental protection from wealthy state residents, and proceeds from state lotteries which are tied to environmental activities.

The variety of fees that the states have developed to pay for environmental programs is staggering. For example, in Wisconsin 32 types of fees are in place including fees on the purchase of new cars; acid rain fees for the issuance of permits for air pollution control; fees for drilling wells, for emplacing storage tanks, and for installing septic systems; and fees for the sale of pesticides, fertilizers, and soil and plant additives.

Other states are also ingenious in finding ways to finance environmental programs. Washington has a "sin tax" on tobacco and cigarettes to finance its water quality programs. California's air pollution programs are financed partially by taxes on vehicle registrations and drivers' licenses. Many states have found that general obligation bonds and revenue bonds are lucrative ways to support environmental programs. However, bond issues require legislative or voter approval, approval which may be difficult to obtain.[5]

Financial constraints have not prevented the enactment of many types of state programs designed to curb chemical threats to the environment. New Jersey, for example, has established four programs to reduce human health risks from accidents involving toxic chemicals. The Toxic Catastrophe Prevention Act stringently regulates manufacturing and other facilities handling 11 chemicals which have been identified as extraordinarily hazardous. The state can order any of the facilities involved with these chemicals to make manufacturing design modifications if there is even a remote possibility that an accident could occur. The Worker and Community Right to Know Act covers 30,000 facilities ranging from manufacturing complexes to hospitals and auto repair shops. This act requires employers to clearly label and report to the state hazardous substances at their places of business and to provide detailed safety information to their workers. New Jersey's incident reporting law requires a company to immediately notify the state of any accidental chemical release which could cause off-site problems, including off-site odors. Finally, the state trains 3000 emergency workers each year to detect and respond to chemical hazard problems.[6]

Reflecting the regulatory bite in its environmental efforts, the state boasts that its approach allows the administrators of its hazardous waste program, for example, to take action to assess and clean up hazardous sites without waiting for recalcitrant or bankrupt dumpers to respond to state requirements. At the same time, New Jersey's enforcement arm assesses penalties and litigates for injunctive relief as necessary. All the while the responsible parties accrue broader liability under the state's Spill Compensation and Control Act, Water Pollution Control Act, or Solid Waste Management Act.[7]

Another signal of increased state sensitivity to public health and ecological concerns is the establishment of the position of environmental advocate. Such an official is already in place in New Jersey, and a similar position is being considered in California. An advocate in Sacramento would have the authority to file lawsuits against violators of pollution regulations and would also administer a $40 million environmental research program.[8]

Even if the federal emphasis on environmental protection intensifies, many states will continue to press forward with their increasing ecological consciousness. They will develop new programs as individual states, and increasingly we can expect to see coalitions of states emerge to coordinate their efforts for environmental reforms in Washington. The efforts of the states to maintain their fiscal and regulatory integrity are reflected in the following set of principles developed by a coalition of state legislatures:

> (1) that the federal government minimize the extent to which it mandates state laws or regulations without providing adequate funding to support the program: (2) that the federal government resist the temptation to preempt state laws; and (3) that Congress pass no legislation and the Administration adopt no regulations that violate the integrity of the intergovernmental fiscal system.[4]

In summary, the states want to be a dominant force in environmental protection activities, with Washington providing as large a portion of the needed financial support as possible. They want to establish their own regulations, and they want few strings attached to federal funds. At the same time, they realize the importance of a degree of national consistency in environmental protection regulations, consistency that can only be brought about by federal leadership and federal regulations. Further, they recognize the impressive concentrations of technical experts within the federal agencies who can be very helpful in supporting their efforts, and they want to tap these resources. Finally, they realize that sometimes local political realities thwart scientific rationality in addressing controversial issues and that interventions by the technical agencies of the federal government can be helpful in fending off politically motivated programs which could be damaging to the environment.

Now let us turn to the problems of the runoff of contaminated water and the control of pesticides to illustrate the roles which the states are playing and the struggles among federal, state, and local authorities for dominance in environmental protection.

Recognizing the Impacts of Chemical Runoff

For many years, the regulatory emphasis in Washington and in the state capitals in protecting the quality of the nation's water resources has been on mitigating pollution from industrial facilities, from municipal sewage systems, and from other easily recognizable discrete sources of contaminants. Such sources are known as "point" sources of pollution, and they have been responsible for much of the toxic loading placed on the environment. The Clean Water Act defines a point source as ". . . any discernible, confined, and discrete conveyance, including but not limited to any pipe, ditch, channel, tunnel, conduit, well, discrete fissure, container, rolling stock, concentrated animal feeding operation, or vessel or other floating craft, from which pollutants are or may be discharged."[9]

While continued efforts to reduce the volume of chemicals flowing from point sources into groundwater and waterways remain of great importance, the nation needs more emphasis on controlling runoff from urban areas, from agricultural lands, from timber tracts, and from mining regions. As rainstorms wash city streets, chemical debris ends up in nearby streams and rivers. As erosion of topsoil intensifies, soil particles laden with agricultural chemicals find their way into surrounding marshlands and lakes. As melting snow drains over piles of slag from coal mines and other mineral extraction activities, the countryside becomes impregnated with toxic metals. Federal laws have set the framework for reducing such pollution—called "non-point" source pollution. Still, the major burden of the effort to contain runoff rests with state, county, and local governments.

At present, for example, almost all iron pollution in streams, lakes, and estuaries can be traced to nonpoint sources. More than one-half of the zinc and lead showing up in waterways comes from runoff from cities and mining areas. Agriculture runoff is responsible for more than one-half of the nitrate pollution in streams and lakes. While under normal conditions small amounts of these contaminants are not harmful, high levels of lead and nitrates have been tied to disorders of the nervous system and blood diseases, respectively, and iron and zinc can impart bitterness to water supplies making them unpalatable. Many more harmful materials—such as asphalt, solvents, arsenic, and mercury—also drain through complicated pathways before reaching the nation's waters. Chemical runoff can be particularly damaging to groundwater which feeds drinking supplies, and in some areas the costs of providing clean drinking water from either surface or groundwater are directly linked to the level of water contamination from runoff.

Turning first to agriculture, governmental approaches to control runoff of pesticides, fertilizers, and other chemicals have traditionally relied on voluntary programs undertaken by farmers with the federal government sharing the costs. A number of these programs are directed to reducing soil erosion which certainly helps stem chemical runoff. However, chemicals also flow off-site through irrigation systems, and on-site they may percolate straight through the soil into groundwater.

Many state governments now believe that expanded regulations must supplement the voluntary programs of the past in limiting runoff of agricultural chemicals. Farmers frequently object to such a change. Among the regulatory approaches being advocated are prohibitions on the use of chemicals near water wells, tax codes which encourage environmentally sound farm management practices, and prohibitions on significant transfer of chemical pollutants off-site.

Since storm sewer systems and paved streets are among the major conduits for urban runoff, local governments are the ones responsible for containing nonpoint pollution in industrial areas. Unlike farmers who have become accustomed to responding to voluntary-incentive programs, most urban dwellers expect regulatory controls to limit pollution in metropolitan areas. Clearly programs to control runoff should be closely linked to local planning and zoning activities, with stormwater controls in particular intimately tied to zoning. Unfortunately, retrofitting drainage systems in areas where planning had been inadequate and where uncontrolled runoff is now a common problem is often very expensive. Occasionally, low-cost techniques such as small catchment basins can help. In any event, significant expenditures are frequently essential to avoid ecological disruptions and maintain clean water.

Local programs to control land use should encompass the water impacts of small-scale residential developments as well as the problems associated with large-scale industrial developments. While a single housing tract may not seem to pose an environmental problem, the cumulative effect of runoff from many small tracts can be a major determinant of the flow and quality of surface waters over large areas.

Tax breaks and other financial incentives can often be used by states and localities to encourage private individuals to protect land and water resources through constrained development. Local governments can recognize and encourage the efforts of conservation groups which stimulate community actions to limit the flows of chemicals reaching parklands or preserves. Also, government agencies should consider setting aside as protected areas regions that are particularly sensitive to ecological impacts. Minnesota, for example, intends to use a portion of its revenues from the state lottery to purchase important ecological habitats as areas to be protected from economic exploitation.

Since many of the timber regions of the country are under federal or state control, large programs for preventing chemical runoff rest squarely in the hands of government agencies. The intensity of timber operations varies radically among and within stages. However, in every case long-range planning which accommodates approaches such as streamside buffer zones and carefully designed access roads is critical. Such planning should take into account both short-term and cumulative environmental impacts and should provide a basis for ensuring the adequacy of water-quality and other resource protection measures.

Timber areas may be huge and exploited by many timber operators. Clear guidelines for these operators should constrain their lumbering from having adverse impacts on both surface and groundwater. Enforcement is critical since individual lumbering operations which violate established environmental guidelines can sometimes damage an entire watershed.[10]

Mining often creates severe water-quality problems. In some states, such as Montana and Pennsylvania, pollution from mining is a major cause of degradation of water resources. Generally the most severe mining pollution in a region is limited to a few locations near the mines, and some of the potential sources of runoff leading directly from the mining operations can be controlled as point sources. However, when mine spoils are scattered over the countryside, point source controls may be inadequate.

Also, abandoned mines often cause major problems. Ideally, plans for containing runoff from an abandoned mine should be developed before the mine is opened. Unfortunately, many abandoned mines, with no identifiable responsible parties, are currently significant pollution sources, and government agencies may be forced to establish reclamation funds to clean up abandoned mines.

Regardless of the types of programs created to control chemical runoff, the costs will be substantial, and the question of funding will be a core issue. In the agricultural sector, farmers have become used to receiving federal funds to help address their problems. Now, the environmental demands are far outweighing any increases in available funds. With regard to urban areas, the federal government sometimes provides limited funds, generally in the form of grants for planning and assessment. Some states have sizable appropriations to add to this sum, although other states spend less for nonpoint pollution control statewide than they do for a single sewage treatment plant.

In general, programs to combat chemical runoff cannot depend solely on general revenues available to the federal or state governments. The nation should move increasingly toward a principle that the polluter pays. Such a philosophy should be both a deterrent to polluters and a source of funds for prevention and cleanup. However, such a philosophy will be painful for many to accept, and as we have seen in the Superfund program, implementation will be more complicated than administration of a program of governmental loans and grants.

Underscoring previous suggestions, more attention should be given to taxes, including taxes on agricultural chemicals which will discourage excessive use of fertilizers and pesticides. For those installations that encompass large paved areas which spread rather than collect runoff, a stormwater fee might be appropriate. When land is to be transferred for development, transfer taxes which provide revenues to help cushion environmental impacts should be considered. Also, timber and mineral assessment taxes may be needed to address the problems in those sectors.

The problem of runoff is serious in some areas and is becoming more acute as industrial and agricultural development continues throughout the country. The states are clearly the key players in the transition to more aggressive control of chemical runoff. They need to build on innovative approaches of the past in developing programs which are both affordable and effective.

States Want to Get Tough on Toxics

In 1986 California voters rebelled. In a statewide referendum, they overwhelmingly decided to "get tough on toxics," and they passed

Proposition 65, The Safe Drinking Water and Toxic Enforcement Act. Proposition 65 sent shock waves through the business community in California and in other parts of the country where similar legislation soon became a lively topic of debate in state legislatures. The new law stunned Washington as the federal agencies tried to figure out how an initiative of this magnitude to control toxic chemicals could originate in Sacramento when up until that time the federal government had considered itself to be the pacesetter in developing the ground rules for controlling toxic substances.[11]

This vaguely worded state law prohibits businesses from discharging into groundwater and other sources of drinking water those chemicals "known to the state to cause cancer or reproductive toxicity." Also, it requires businesses to post warning signs before exposing anyone to such chemicals. The law has a "bounty hunter" provision which permits private citizens as well as public enforcement officials to take violators of the law to court, and the vigilant citizens receive one-fourth of any fines that result from the court actions which they initiate.

To the voters of California, these provisions certainly seemed reasonable. They had been inundated with media accounts of the leakages of toxic chemicals into drinking water supplies from many hazardous waste dumps in both southern and northern California. They had learned that even though Silicon Valley was generally free of large smokestacks, the electronics industry was not as clean as advertised as leaking storage tanks of the electronics firms discharged into the groundwater solvents which were to be used in their production processes. Finally, the stories of toxic runoffs from agricultural areas impacting on residential areas and on wildlife preserves were increasing. Symbolizing environmental concerns, one-third of the population of southern California already relied on bottled water, including many who wanted to avoid possible health problems even though the tap water was judged by the health authorities to be perfectly safe.

At the same time, throughout the nation citizens had steadily lost confidence in the integrity of government. In California, voters had become particularly disenchanted over an apparent lack of determination of the federal and state agencies to adequately enforce pollution and waste regulations already on the books. The EPA scandals of the early 1980s revolved in part around several waste sites in California where the citizens had become outraged at the lack of cleanup action. The state agencies were perceived by many residents as spineless.

Thus, Californians were sympathetic to the concept of environmental vigilantes who would ensure that toxic polluters would be punished.

When the law was enacted, the scientific debates throughout the country over what types of chemical discharges cause cancer were intense. Environmental groups argued that hundreds of "probable" carcinogens should be controlled by the state. Their lists included many chemicals which were widely used in agriculture, industry, and dry cleaning establishments. They urged placing severe restrictions on the handling of these chemicals to prevent their reaching the groundwater, restrictions that would cause major economic dislocations. In 1987 the state, sensitive to these economic implications, identified only 29 chemicals for regulation, selecting those chemicals which their scientific advisers concluded had been shown to cause cancer in humans while not including the many other chemicals which were of concern because of their biological effects on laboratory animals.

Also, controversy arose immediately over the amounts of these chemicals that cause harm. Again, some groups argued that no level of discharge of these chemicals above zero was acceptable. Within several years, however, the state settled on controlling only those discharges which produced pollutants above a "threshold" level of concern. This level would be established on the basis of risk estimates that a discharge would cause less than one case of cancer in a population of 100,000 people. As we have seen in previous chapters, the uncertainties in deriving such risk factors are very large. At the same time, however, the state needed specific discharge levels as the basis for regulation and enforcement.

Turning to the threat of reproductive toxicity—or the adverse effects of chemicals on prospective mothers and their unborn children—the state, like the federal government, had considerable difficulty identifying chemicals which should be of concern let alone in determining safe discharge levels. Aided by scientific advisory groups, the state has spent considerable effort to respond to this legislative mandate. In 1990 the governor listed 200 chemicals in this category despite the lack of a solid research base to provide good guidance concerning subtle chemical threats to the process of human reproduction.

With regard to the requirement of Proposition 65 that businesses warn customers of the hazards of their products containing toxic chemicals, the initial predictions that supermarkets and retail outlets would be plastered with warning signs have not materialized. Indeed, the state government promptly determined that drugs and food were beyond the reach of the new law, a determination which was overturned by a California court in 1990. Initially, the most widespread warnings were those placed at every outlet selling alcohol which pointed out the dangers of pregnant women consuming alcohol. Should warning signs now be required for every carcinogen and reproductive toxin found in drugs and food? Also, environmental groups are demanding that warning signs be attached to shelves displaying dozens of brands of paint removers, spot removers, and water repellants containing minute levels of widely used methylene chloride and perchloroethylene. Thus, the character of many stores could change dramatically, particularly if very small traces of naturally occurring toxic chemicals which find their way into food such as aflatoxin in peanuts are also included.

Other states were initially intrigued by California's boldness in assuming a leadership role in addressing toxic chemical problems. A few drafted their own versions of Proposition 65. However, most state legislatures have been reluctant to enact such legislation given the difficulties of translating general regulatory intentions into meaningful programs. The National Governors Association decided not to endorse Proposition 65 as "model legislation." They are wary of the ability of the states to cope with the scientific uncertainties as well as the economic ramifications in controlling large numbers of chemicals.[12]

As repeatedly noted, many states are dissatisfied with the lack of leadership in Washington in providing an adequate basis for controlling toxic chemicals. At the same time, the states recognize their limitations in going it alone, particularly in coping with the scientific uncertainties when trying to determine the details of regulatory approaches. Meanwhile, local politicians worry about how the economies of small communities may be affected by tough new laws. At the local level, there is inevitably a gap between the degree of protection demanded by many citizens and the cost they will bear to achieve this level of safety.

The federal government has the necessary legal authority to address the dangers of toxic chemicals wherever they may appear. Furthermore, the scientific resources available to the federal government are very large. The EPA and the other federal agencies simply need to give a higher priority to working together and working with the states—one by one—in addressing the problems of toxic chemicals. The states are increasingly determined to play a greater role, and the federal government must find better ways to integrate efforts in Washington with those in the state capitals.

Keeping Pesticides Out of Groundwater

For several decades pesticides have been the toxic chemicals of greatest public health concern in rural America, and they deserve more detailed comments. Designed to kill weeds, bugs, or fungi, they also can harm farm workers who have not received proper instruction on how to apply them in the fields. They can cause health problems for children playing in areas in the wake of crop dusters and for rural inhabitants who find pesticide residues in their drinking water wells. Every state is now trying to ensure that such public health threats are minimal without disrupting farming practices.

Still the nation has a long way to go. Every summer as I play tennis in a public park in Arlington, Virginia, a helicopter sprays the immediately adjacent parkland with little regard to those of us using the nine courts. Of course, pesticide drift is far more serious in rural areas of the West. Also, complaints from migrant workers in the San Fernando Valley suggest that they are not always properly trained to handle pesticides. Finally, as analytical chemistry techniques become ever more sensitive, more and more pesticides are being detected in crop, soil, and water samples taken from agricultural areas.

Let's take Florida as an example of a state which has made impressive progress in capping pesticide problems that have threatened water supplies. Agriculture is a $5 billion per year business in the state. Very permeable soils expose a high groundwater table to chemicals percolating downward from the surface. This is particularly important since 90% of the population relies on groundwater as its source of drinking water. In 1982, discovery of the pesticide ethylene dibromide—a carcinogen—in groundwater and in foods in retail outlets galvanized state agencies into action. State officials were determined to reduce groundwater contamination and prevent further degradation of this critical resource.

Limitations on pesticides cut across the interests of several Florida state agencies and affect many economic interests. The three agencies most concerned—those responsible for agriculture, health, and environmental protection—agreed to consolidate their efforts. They began by classifying groundwater areas according to their uses and susceptibility to chemical contamination from the surface. They then expanded monitoring of groundwater across the 67 counties of the state. The state also provided financing for a special fund for cleaning up emergency pesticide situations.

The state established regulations for use of certain pesticides. Some cannot be used within 300 feet of drinking water wells and 1000 feet of other wells in particularly permeable areas. The agencies have tightened procedures for approving pesticide use in the state, and particularly for granting exceptions to general prohibitions on using particularly toxic chemicals. Finally, the number of enforcement inspections has increased dramatically.

Florida officials concede that some groundwater degradation is unavoidable and must be tolerated. The state has been characterized as a "sand bar connected to Georgia" because of its permeable soils. As long as agriculture remains a backbone of the state's economy, some low levels of pesticides will find their way into the aquifers although aggressive management prevention strategies can reduce the severity of this problem. About 20 pesticides are commonly present at trace levels in samples from water wells in a number of areas. Now the task is to prevent the degradation of groundwater to levels of potential health concern. During the past several years, discoveries of samples that are contaminated above the safe levels prescribed in Washington for drinking water have been rare.[13]

As discussed in the previous chapter, a highly detailed national plan for protecting groundwater is not appropriate given the variations in subsurface geology, land uses, and agricultural and industrial uses. Rather, individual state-specific plans are in order. However, aquifers do not respect state borders, and states must learn to work together. Also pesticides can be bought in one state and used in another. Therefore, state limitations on the sale of certain pesticides may not be effective. The EPA can play an important role both in encouraging cooperation among states and in ensuring a "level playing field" for businesses by establishing minimum standards for all states. States which are lax in their protection of groundwater should not reap an economic advantage over more diligent states whose farmers compete in the same markets.

Thus, the EPA should ensure that all states have at least minimal plans for protecting groundwater resources from pesticide contamination. Of course the states must have programs to translate the plans into action and to enforce compliance. In those states which do not respond in this area, the EPA should conduct its own assessments of groundwater problems and undertake the necessary steps to ensure proper management of pesticides. In such states, the EPA might very well cancel the use of certain pesticides throughout the state or in specific counties. Though such actions may raise local protests of federal meddling in state affairs, the environmental message will come through.

In 1988, the EPA Administrator asked:

> We have been questioned about whether this change, shifting pesticide management from primarily a federal function to one more dependent on an increasing management role by the states, is appropriate. Is it realistic? What requirements from the EPA will be needed to assure state action, yet not interfere unduly with each state's ability to tailor its program to its groundwater conditions?[14]

The EPA is on the right track in challenging the states to exert leadership in protecting their water resources. The states need encouragement and support. Then if some states do not respond, a return to federal intervention in those states may be necessary.

The Importance of an Informed Public

In 1980–1982 and again in 1989–1990 infestations of the Mediterranean fruit fly, or medfly, threatened the multibillion dollar fruit and vegetable industries of California. Helicopters sprayed the pesticide malathion widely to help check the spread of this destructive insect. Such spraying had to be targeted directly over urban as well as rural areas in order to be effective. This episode illustrates the central role played by state agencies not only in designing programs for local uses of chemical pesticides but also in informing the public about the levels of hazard associated with exposure to chemicals.

Prior to the 1980s, Malathion had been frequently used for 35 years in farming regions and in residential areas to combat insects throughout the country, apparently without harmful side effects. Still apprehensions among the populations in the helicopter flight paths were understandable. In 1981, my daughter was living directly under one of the helicopter routes in Palo Alto, and I, like many others, quickly reviewed the toxicity information concerning malathion.

At that time, the California Department of Food and Agriculture went to extraordinary lengths to reassure the population that the spraying would be deadly for the insects but harmless for humans. In special leaflets and through the media, California state officials widely announced the location and timing of the helicopter sorties into urban areas. The department pointed out that the spray would be about as toxic as laundry detergent. The message seemed clear. Stay indoors so you don't breath the new type of "detergent," but don't worry about touching it.

The department was quite thorough in its consideration of possible problems and in the information it provided to California residents. It noted that there was little danger to humans including children who played in sprayed areas. However, cats and fish and even automobiles could be at some risk. The mixture sticks to fur. Since cats groom themselves, they are more likely to ingest malathion. Fish in very small ponds could be at risk since a small amount of malathion concentrated in a very limited area might have a disruptive effect on their biological balance. As for automobiles, the chemical can be tough on paint and should be promptly washed off.

Of course the spraying took place late at night when most people were inside. Hospitals were avoided to the extent possible. Follow-up studies from the 1981–1982 sprayings, including one study of 6000 pregnant women in the flight paths, revealed no short-term or long-term effects from their limited exposure, if any, to malathion.

Then, in 1990 the spraying was extended to Los Angeles where anxieties and emotions of the public run even higher. Some scientists, environmental groups, and public critics of the program challenged the safety assertions of the state, pointing to new research reports of the breakdown of chromosomes associated with malathion exposures. While the state continued to champion the safety of malathion, it became sensitive to these concerns and convened additional scientific panels to examine the program once again. Scientific judgments may differ concerning the potential public health problems associated with the program, but the state must be given high marks for its efforts to inform the public of the basis for its decisions and precautions that should be taken in this controversial area.[15]

Another example from California illustrates the growing importance of state and local governments in convincing the public of environmental costs and environmental benefits from actions or inactions. In this case the objective was to encourage the public to accept its responsibility to clean up toxic wastes. In 1986 and again in 1989 Marin County north of San Francisco sponsored Household Hazardous Waste Collection Days. In four hours about 1200 cars deposited enough toxic waste outside the convention center to fill several large trailor trucks. The bulk of the waste was paint. Also large quantities of pesticides, paint thinners, and other solvents were included. Some brought waste oil, remnants of asbestos, old furniture polish, caked bathroom cleaner, and unwanted insect repellants.[16]

While 1200 households are but a small fraction of the total population of the county, a start has been made, and the widespread publicity has certainly raised toxic awareness. Meanwhile, the county has provided every household with easily understood guidelines on handling household toxics. Printed on a chart which can hang in the garage, the guidelines advise how to handle hazardous materials at home—how to store them, how to dispose of them, and how to find less toxic, alternative products.

In the same educational vein, Wisconsin has completed an important project to acquaint its citizens with the concept of "risk." In cooperation with the University of Wisconsin, the Department of Natural Resources has produced an excellent brochure discussing in laymen's terms most of the factors commonly debated when considering environmental problems. Under the heading "You can learn a lot from a rat," the authors quickly add, "but not everything." They proceed to explore the complexities of toxicology in a way that develops realistic perspectives of the role of science in making risk decisions. The brochure identifies 25 of the most common chemical risks encountered in the home and outside the home, with tips on how to avoid these risks and telephone numbers to obtain more information.[17]

Many other states are also launching major programs to help the public better understand the chemical environment they live in. Arizona has videotapes, films, and slide shows available on environmental issues. They provide handouts and teaching materials upon request. They even suggest tours of facilities engaged in chemical activities and guest speakers to address common problems encountered throughout the state. Louisiana is preparing course materials on 100 environmental topics and has published a three-volume teaching guide. Ohio requires their school curriculum to include the concept of conserving natural resources. Its Department of Natural Resources has established an Adopt-a-School program for which it provides extensive course material to the participating schools. Virginia has a Chesapeake Bay teacher project which provides opportunities for teachers and students to visit the Bay. The state also sponsors environmental courses for teachers at four universities during the summer.

Clearly, education is a principal key to better approaches to environmental protection in the years and decades ahead. The states are gradually accepting their responsibilities in environmental education, and the federal government should strongly support such efforts—with both encouragement and financial resources.

The States Prepare for the Long Haul

State governments have come a long way since the first Earth Day 20 years ago. They have established environmental agencies, adopted wide-ranging policies and programs to curb the spread of environmental chemicals, and supported educational programs which sensitize the public to environmental risks. Environmental issues are now central in the platforms of almost every state politician, and environmental lobbyists are firmly entrenched in every state capital. The governors recognize that "when the people have a hazardous dump in their backyard, they don't call the White House, they call the statehouse." Meanwhile, the states continue to battle in Washington for larger shares of federal funds appropriated for environmental protection. Counties and cities have entered the fray in the state capitals for greater control over both federal and state funds.

From industries which must obtain state air and water discharge permits to households that must segregate toxic trash, regulated parties have no choice but to pay greater attention to the environmental programs of the statehouses. As California prepares to require new cars to have computerized devices that warn drivers of malfunctions in exhaust systems emitting air pollutants and as Tennessee strips local communities of their right to veto hazardous waste storage and treatment facilities in their neighborhoods, the role of state environmental agencies is growing steadily.

In the early 1970s, I visited senior environmental officials in Sacramento, Madison, Annapolis, and Columbus, seeking their support for stronger action in Washington and in the states to control toxic chemicals. At that time they had little incentive to worry about trace levels of chemicals, toxicity, or exposure—topics which were on center stage in Washington. Instead, they were preoccupied with smoke hovering in dirty skies, lakes and rivers choked by inadequately treated sewage, and rat-infested garbage dumps. They were glad that the federal government was handling the problem of trace chemicals in the environment since they had neither the time nor the manpower to devote to these secondary problems.

In the late 1980s, I visited senior environmental officials in St. Paul, Hartford, Annapolis, and Denver. What a difference! Priorities had changed as much of the highly visible pollution had been curbed and as public sensitivity over toxicity had emerged. The states were responding to the toxic challenge. Highly trained specialists were in place. They had an excellent appreciation of the spread of chemicals in the environment—the sources and the sinks. They were glad that the federal government was also taking steps to check the spread of toxic substances, insofar as these efforts were designed to support the efforts at the state level. Meanwhile, they repeatedly complained that Washington was so bogged down in developing theoretical concepts and vague strategies that the federal government simply was not keeping up with the real problems of chemical contamination confronting the states.

Speaking for his state, the governor of Colorado recently said: "We seek neither benign neglect nor federal tutelage. What we seek is a pragmatic partnership with the federal government . . . understanding this nation's diversity will be the key to environmental progress." Noting that while the states must play a lead role in resolving their own problems, he emphasized that the states cannot do it alone. As a specific example, he added, "Federal deadlines are useful because they pressure us to take action to improve the air. But they mean nothing if we are not given the tools to succeed."[18]

A continuing tug-of-war between Washington and the states as to who will set environmental policies, who will design and control action programs, and who will distribute public funds is inevitable. In many ways this tension which keeps all parties on their toes reflects a healthy relationship. At the same time, the regulatory muscle for controlling toxic chemicals, which used to be concentrated almost exclusively in Washington, is rapidly shifting to the state capitals. The EPA and other federal agencies must now recognize that the states intend to take their environmental responsibilities more seriously than ever before as described in the mission statement of the Minnesota Pollution Control Agency:

> . . . to serve the public in the protection and improvement of Minnesota's air, water, and land resources by: assessing the state's environmental status; regulating the quality of these resources; assisting local government, industry, and individuals in meeting their environmental responsibilities; and implementing strategies that will protect and enhance public health and the state's environment.[19]

The federal government should welcome this hardening of the environmental muscles of the states, for they are on the firing line. They need and they deserve support from Washington—support that reinforces the saying, "Think nationally, but act locally." Protecting a community's environmental resources is synonymous with protecting the resources of the nation.

7 The Greening of Industry

Pollution is nothing more but resources we're not harvesting.
—Buckminster Fuller

The biggest challenge we have is convincing the public that we are their friends and not their enemies. . . . Everything can be cleaned up and managed in a very acceptable way with respect to the environment and the health of our people.
—Chief Executive Officer of the DuPont Company Richard Heckert

A Large Company Looks Ahead but Stumbles with the Present

I first visited the headquarters of the 3M Company in St. Paul, Minnesota, in 1975. Best known to most Americans as the manufacturer of Scotch Tape and Scotch Gard, the 3M Company has for many years been an industrial leader in the production of many types of adhesives and coatings for industrial and consumer uses. The company also manufactures electronic circuits, pharmaceuticals, audio and video goods, and a variety of other products at facilities in 24 states and 22 foreign countries.

The company's environmental coordinator was pleased to receive me as a representative of the EPA. He did not hesitate to boast of the company's major investment in the construction of a new high-temperature incinerator. Using this new facility, the 3M Company could destroy many of the toxic wastes generated by its plants. In addition, the company could offer incineration services to other firms that were also seeking better ways of disposing of these wastes.

This facility was coming on-line at a time when industry was seized with a need, nationwide, to destroy hazardous chemical wastes rather than place them in landfills where some would remain toxic for centuries. At that time the temperatures being reached at most of the incinerators throughout the country were too low for such destruction and allowed many unburned chemical products to escape into the environment. The 3M incinerator would avoid this problem.

While the incinerator was too new to have demonstrated its full potential, 3M officials were confident that this state-of-the-art device would solve many of their waste disposal problems. In addition, the publicity which they had disseminated about the facility was receiving very favorable reviews within Minnesota and in Washington. The company was rightfully claiming national leadership in the field of high-temperature incineration, for its incinerator was at the frontier of technology.

The 3M management had also developed another program which was in the early stages of implementation. The concept behind this new program called Pollution Prevention Pays, or 3P, was very simple. If individual employees could develop ideas to modify existing manufacturing approaches or could design new approaches which made sense economically and at the same time reduced the amount of wastes being generated, the 3M Company would give them special recognition. The company encouraged every employee to undertake this task, and a special review board of senior company managers evaluated the ideas presented by the employees.

When I returned to the 3M headquarters building 14 years later, I received a glowing report on the tremendous success of Pollution Prevention Pays, but I did not hear a word about the incinerator. By 1989, "waste minimization" as an alternative to waste disposal had become the most highly promoted slogan in Washington, and 3M had been the first manufacturing company to formally embrace the concept through its 3P program. Indeed, representatives of many other companies regularly visited 3M headquarters to learn about the program and to pattern their approaches after the 3P method.

According to officials of the 3M Company during my second visit, 700 projects under the 3P program had resulted in an annual savings of over $400 million for company operations in the United States alone. At the same time, they pointed out that over the years these projects had prevented the discharge into the environment of 11,000 tons of air pollutants, 15,000 tons of water pollutants, one billion gallons of wastewater, and 388,000 tons of sludge and solid wastes. The changes had been made in four general categories: changes in the chemical formulations of products, modifications of production processes, redesign of manufacturing equipment, and recycling of materials which could be reused in production processes. Corporate management gave special recognition to projects which were innovative, incorporated original designs, or involved significant technical achievements.

Undoubtedly, many if not most of the projects would have evolved in the absence of the 3P program as the company tried to save money through improved technologies and complied with federal and state regulations limiting the discharges of pollutants. However, some of the approaches probably would not have been developed in the absence of the constant pushing of the 3P program by senior company officials. The program has also provided the 3M Company with a very impressive scorecard of concrete steps that have been taken on the initiative of 3M employees to reduce environmental contamination.

During my second visit the incinerator was not a topic of discussion for an understandable reason. Prior to my visit, I learned from the Minnesota Pollution Control Agency that the agency had just fined the 3M Company $1 million for improper operation of the incinerator. These faulty procedures apparently had allowed trace levels of pollutants considered hazardous by the state to escape into the atmosphere for many months or even longer. I can only surmise that company management had become overly confident in the advanced technology and did not pay enough attention to the human factors involved in day-to-day operations.

Also, I learned that 3M officials had concluded that in recent years they had not given adequate attention to the reductions of chemical emissions into the atmosphere at the company's manufacturing facilities, reductions beyond those called for in governmental permits. Many facilities, according to the officials, had been sited in relatively remote areas where air emissions presumably would not cause concern to the public, and therefore the company had assumed that additional steps would not be necessary.

However, in the mid-1980s the attitudes of federal and state environmental agencies, and of course the public, toward emissions of chemicals into the atmosphere regardless of the locations of the facilities had changed dramatically. Nationwide concern over toxic air emissions heightened when the EPA, in response to newly enacted legislation which had been supported by industrial lobbyists, required all companies to prepare reports for public release of their total emissions of a long list of those chemicals of greatest concern on a plant-by-plant and chemical-by-chemical basis. 3M, for example, reported that in 1987 its 50 plants in the United States had discharged 61.7 million pounds of these chemicals into the atmosphere. The company officials were quick to point out that in view of the broad diffusion of these emissions the resulting concentrations in populated areas were extremely low, and indeed seldom detectable.

Like many firms in 1989, 3M was responding to the pressures of the regulatory agencies and the public to reduce discharges of hazardous air emissions, as well as other types of emissions, into the environment. Just before my second visit to St. Paul, the company had announced its new 3P *plus* program. This concept adds specific pollution reduction targets to the 3P program with line managers responsible for seeing that the targets are met. Specifically, in 1989 the company committed to reducing discharges into the atmosphere by 90% within the 1990s through a combination of air pollution control equipment, substitutes for petroleum-based solvents, and greater emphasis on recovery and recycling of waste products. The company was confident it could achieve this ambitious goal since the company had made considerable progress in fulfilling a 1987 commitment to reduce air emissions by 70% by 1993.

The foresightedness of 3M in adopting the 3P program has been widely commended. Yet the same company stumbled with the routine operation of an incinerator. Also, the company is now recognizing that there are no environmentally "remote" locations in the United States, and even a handful of local residents deserve as much protection from air emissions as do large numbers of city dwellers.

Many other companies have environmental profiles comparable to 3M. Their programs for protecting the environment have many commendable features often pressing the state-of-the-art of pollution control to levels which government agencies had thought could not be attained. Still, almost every major company has soft spots somewhere in its environmental activities which deserve more concerted attention by the company in the years ahead.

Environmental Outrage Engulfs a Small Company

Forty miles south of St. Paul lies Northfield, Minnesota. This rural community has been known to many as the home of Carleton College and St. Olaf College. The informal town slogan has been "Cows, Colleges, and Contentment." However, when I visited Northfield in late 1989, I was greeted by T-shirts picturing a dead cow lying on its back with its feet extended upward and a slogan reading "Cows, Colleges, and Carcinogens." The following words were on the back of the T-shirts:

> Sheldahl Inc., of Northfield, Mn., was among the worst in the nation at releasing known and suspected cancer-causing chemicals into the air in 1987. Associated Press.

Sheldahl was the largest major employer in Northfield with 950 workers at its facilities just north of the town. The company also had plants at three other locations in the United States and had financial ties to a large Japanese firm. For more than two decades the company had manufactured flexible electronic circuits for the communications, automotive, and aerospace industries at its Northfield facilities.

The company's manufacturing processes depended heavily on the use of solvents with carcinogenic tendencies. Other chemicals with particularly desirable electrical properties were dissolved into these liquid solvents. The solutions were washed onto metallic or other surfaces where the added chemicals remained permanently as a result of chemical interactions. The solvents were then either discarded (for example, boiled off the surfaces and exhausted out a stack) or captured and recycled to the extent possible.

According to company officials, over the years Sheldahl had relied on solvents which it considered to be not only technically and economically acceptable but also safe within the production plant and benign in the environment outside the plant. However, as concerns over the hazards of using solvents had increased during the early 1980s in

Washington and throughout the country, the number of solvents considered to be safe dwindled. The company considered methylene chloride to be the most environmentally acceptable of all the solvents which could be used in the process of laminating circuit boards. Methylene chloride has a particularly desirable feature of nonflammability. In some of the company's other processes, only flammable solvents were judged to be technically suitable, and special care was taken in handling these chemicals.

As indicated on the T-shirts, in the mid-1980s Northfield appeared on the EPA's list of cities with significant discharges of carcinogens, and specifically methylene chloride vapors. This chemical became the center of a controversy which penetrated every home in the community. This controversy was similar to clashes over toxic chemicals in other communities throughout the country which erupted when the EPA list of discharges was released.

The risks to the community from emissions from the plant were unknown. Methylene chloride administered to laboratory animals (through ingestion but not through breathing since ingestion experiments are much easier to conduct) at very high levels had caused tumors in some species but not in others. There was no scientific evidence that workers who breathed the chemical had or had not suffered long-term ill effects. The levels of the chemical present outside the plant boundaries were estimated by computer models using questionable assumptions as to the rates of discharge and the behavior in the atmosphere of methylene chloride, and there were no reliable monitoring data available to determine the levels of the chemical in the air within or outside the facilities. The company had reported to the EPA that a certain quantity of methylene chloride was used up in the production process, and it was simply assumed that most of the used materials escaped into the atmosphere.

Faculty members at Carleton College had examined the potential risk. They released an assessment of the hazard prepared by a student who had a very limited understanding of the concept of uncertainty, a concept which is inevitably involved in risk assessments. His poorly documented paper simply added to the controversy and at least temporarily weakened the credibility of the college scientists.

Many townspeople argued that any risk was unacceptable and that methylene chloride should no longer be used in the plant. Other residents worried about their jobs and urged careful consideration of the economic implications of overreacting. The company had already agreed with the local labor union to reduce the quantity of methylene chloride being used by 40% within one year. The company argued that economically it was impossible to phase out the chemical completely for at least four years and suggested that in the interim the emissions be diluted at ground level by using a higher exhaust stack.

Meanwhile, the Minnesota Pollution Control Agency was struggling with the enforcement of its requirement that a "maximally exposed" person standing outside the plant would not be subjected to a risk of more than 1 in 100,000 chances of receiving cancer from exposure to the chemical. One popular interpretation of "maximally exposed" was that an individual would remain for an entire lifetime at the point of maximum concentration of the chemical along the fence of the plant. The state had great difficulty determining what that concentration should be. Everyone was scrambling to become quickly educated on the intricacies of risk assessment only to learn that science has severe limitations.

While the debate over the future of the Sheldahl plant continued, several conclusions from this and similar situations around the country seemed clear. The congressional requirement that all manufacturing facilities publicly disclose the amounts of chemicals being discharged into the environment has had a dramatic effect. It has forced companies to accelerate their plans, and in some cases first develop plans, to reduce chemical discharges. State authorities have awakened to their past neglect of chemical pollution problems. Individual citizens are beginning to realize that for many years they have been surrounded with the chemical by-products of an industrial society. Finally, most Americans seem to be reacting to these new revelations the same way they have reacted to smog: let's eliminate the chemical pollution if we can, but let's not worry about it if we can't. However, a few are more insistent: discharges must stop now even if we must shut down the facilities.

We should not be overly critical of the Sheldahl Company's neglect of chemical emissions in the past. After all, for many years state authorities had not been concerned. Small companies cannot be expected to have the array of experts available to the government. American industry had become accustomed to reacting to governmental requirements. Voluntary actions on the part of industry have not been a high priority, either by industry or by government—at least until now.

The Sheldahl Company along with many other much larger companies relied on methylene chloride as the most technically and environmentally acceptable solvent, and indeed it was preferable to many other possible choices. Understandably, Sheldahl sought to solve the problem in a way which would be least disruptive economically. Finally, there is considerable merit in the company's argument that the uncertainties associated with the risk estimates are so great that the debate over the *degree* of risk had become almost meaningless.

In the future, however, neither Sheldahl nor any other company can hide behind the excuse of not being aware of the possible environmental problems associated with chemical discharges into the air, into the water, or onto the land. Through the requirement for public disclosures of their discharges, all segments of American industry have been put on notice that they are expected not only by governmental authorities, but also by the American people, to reduce chemical discharges. Arguments will persist over the rate and extent of these reductions. But regardless of the inability of environmental advocates to demonstrate specific levels of risk, the commitment of American society to reduce exposures to toxics, however small, seems clear.

The Environmental Consciousness of the Chemical Industry

The attitudes and behavior of American chemical companies toward protection of the environment are dramatically different in 1990 than they were in the 1970s when the initial laws to control toxic chemicals were enacted in Washington and in many state capitals. Twenty years ago, aside from restrictions on the use of pharmaceuticals and pesticides, few regulatory barriers inhibited the development and sale of chemicals. Few companies took time to look beyond their internal staffs for advice on whether chemical products were safe since doing so might slow down their marketing activities. Now, judgments of many boards of directors and company managers on the values of products are based equally on marketability and environmental acceptability. This acceptability is defined by governmental agencies, by national scientific organizations, and by a greatly increased cadre of company environmental specialists.

Economic considerations always motivate private companies, and industrial managers now clearly understand that the costs of correcting mistakes from inadequate attention to environmental protection can be devastating to a company's profit margin. The costs to Union Carbide for the Bhopal accident, to Exxon for the Alaskan oil spill, and to Johns Manville as the target of the asbestos liability suits, for example, have been widely publicized. However, almost every company has faced less dramatic yet still substantial costs in coping with environmental problems. These problems have made lasting impressions on the approaches of American industrial leaders to the calculation of likely profits and losses associated with environmental considerations.

The Superfund legislation and related federal and state regulations together with a number of court decisions have clearly established the principle that a company which manufactures a chemical retains some responsibility for ensuring that the chemical does not harm humans or the environment—even if the chemical is sold to another company. This cradle-to-grave liability concept has sensitized all firms to the need for great caution in selecting processors, distributors, and disposers of their chemical products and their wastes and in ensuring that the recipients of their chemicals are committed to responsible environmental procedures. Similarly, some companies are now very wary of the practices of their suppliers of chemicals, lest the suppliers skirt environmental regulations and draw their purchasers into legal entanglements.

As would be expected, many companies protect themselves with elaborate insurance arrangements to cover financial difficulties resulting from environmental problems. Large corporations tend to opt for self-insurance, often setting up separate insurance entities within the corporate structures. Other companies seek protection through the insurance industry which has been doing a land-office business in this area. The insurance companies in turn may require manufacturers to take certain steps to reduce the likelihood of environmental problems. Then they adjust their rates in accordance with their assessments of the degree of risk involved.

Does this mean that all companies are now taking all possible steps to ensure that toxic chemicals will not harm the environment? Of course not. Every week companies are still being fined by governmental authorities for failure to comply with environmental regulations. The data released by the EPA in 1989 indicated that large amounts of potentially toxic chemicals were being discharged into the environment. These data triggered the Sheldahl case and also led to the 3P plus program at 3M. They provide dramatic evidence that much more aggressive action needs to be taken by industry to reduce chemical discharges. Meanwhile, the continuing controversies over remediation of old waste sites often reflect recalcitrance on the part of many companies to clean up their sins of the past. This recalcitrance is often prompted or accentuated by difficulties in resolving liability issues among the manufacturing and the insurance companies which are involved. Still, the long-term costs of not complying with environmental regulations, including the intangible costs associated with tarnished corporate images among the public, are on the rise, and companies are less likely than in the past to seek ways to delay or avoid compliance.

Several types of pressures drive chemical companies toward more assertive behavior to embrace environmental protection. Many laws, regulations, and facility permits are in place to help define the limits of acceptable activities. Second, companies are concerned over liability suits or other citizen actions demanding compensation—by individual workers or customers who claim harm from coming into contact with chemicals, or by environmental groups or labor organizations which represent interests that extend beyond a single individual. Many companies, and particularly companies with product lines which emphasize consumer goods, such as Procter and Gamble, value highly their reputations among the American public. They do not want to be linked in any way with environmental problems which might damage their images and give competitors the slightest edge. Of course, within every company there are individuals at the board level, in management positions, and among workers who have strong personal commitments to environmental protection. They are capable of exerting pressure on companies to become environmentally responsible.

In the mid-1970s, as the EPA official responsible for assessing the environmental policies of chemical companies, I had a unique vantage point to observe the behavior of these companies firsthand. At that time it seemed clear that the government needed to stimulate and support efforts of companies to initiate environmentally responsible approaches. A few companies had assembled large environmental staffs to examine every facet of their activities, and they had strong capabilities to develop far better solutions to their own problems than solutions dictated by the government. However, most companies were reluctant to divert financial resources to environmental activities in the absence of clear signals from the government as to steps they should take.

I soon became an advocate within the EPA for officially recognizing voluntary steps taken by industrial firms to protect the environment. Some companies had mounted elaborate programs for testing the environmental behavior of chemicals years before the EPA had legal authority requiring industry to carry out such tests. Other companies had developed environmental training programs for the employees of customers and suppliers long before the government had given any thought to such responsibilities. Still other companies had voluntarily pulled products off the market and suffered considerable economic penalties as they tried to develop safer products. Finally, some companies had made major scientific contributions toward developing a variety of modeling, monitoring, and toxicology tools for assessing environmental problems on a national basis.

The EPA lawyers were shocked at such an outrageous proposal—to commend a legal adversary. They argued that any acknowledgment that a company was making useful contributions to environmental protection would surely weaken the EPA's case if that company ever stepped out of line and, for example, violated the Agency's permit requirements at one of the company's facilities. Besides, they added, such an approach would send the wrong signal to the Congress, to environmental groups, and to the American public that the EPA was being duped by industry to believe that self-policing was realistic.

The lawyers carried the day. The government's approach to environmental protection was to be based on adversarial confrontation. The EPA would continue to work through the *Federal Register*, through legally binding discharge permits, and through the courts as necessary to require industry to respond to a command and control system of environmental protection. Should industry decide to take steps beyond those that were required, such steps would be welcomed—but they would not count on the EPA's scorecard. The EPA's scorecard would only reflect legal transgressions.

Not surprisingly, industry's response has largely been to do what is required or what may be required in the near future. However, the EPA lawyers were wrong. Industry can do a lot more, and official recognition in Washington of responsible corporate behavior could be an important inducement for a greater environmental consciousness throughout the industrial world. Positive reinforcement can work. Just because you commend a company for certain steps doesn't mean you ignore transgressions it may commit.

Industry Efforts to Reduce Toxic Wastes

"Minimization" of toxic wastes is one of the most popular topics today in environmental circles and is one area where the views of government and industry are converging. If industry produces less waste, America will encounter fewer pollution problems.

The Congress, the EPA, state legislators, and state environmental agencies have given waste reduction top priority. As illustrated by the 3M experience, many industrial firms believe that they can improve their competitiveness by giving greater attention to the technical opportunities for reducing wastes, particularly in view of the ever-increasing costs of waste disposal. Of equal importance, all companies now recognize that they have no choice but to respond to environmental concerns and to adopt a waste minimization attitude.

Federal law requires each company to sign the following certification whenever it ships solid hazardous waste off company property to a storage, treatment, or disposal facility:

> If I am a large quantity generator, I certify that I have a program in place to reduce the volume and toxicity of waste generated to the degree I have determined to be economically practicable and that I have selected the practicable method of treatment, storage, or disposal currently available to me which minimizes the present and future threat to human health and the environment; or if I am a small quantity generator, I have made a good faith effort to minimize my waste generation and select the best waste management method that is available to me and that I can afford.[1]

All companies which ship waste off-site must periodically report to the EPA their efforts to reduce the volume and toxicity of waste generated and to compare their current volume and toxicity of waste with waste produced in previous years. In addition, some states require that companies which do not ship waste off-site but store or dispose of the waste themselves must submit comparable reports. Thus, the specific details of waste reduction programs are in the hands of the firms. While these programs are to be under constant supervision of government agencies, many government officials bemoan their inability to confirm the veracity of industry declarations. Waste minimization should become a cornerstone of future environmental protection efforts. If this goal is to be achieved, positive industrial actions must erode the historical attitudes of mistrust within government which currently inhibit a partnership between government and industry to jointly promote and carry out a most attractive concept.

"Waste minimization" as a legal concept was first articulated in law in 1984 although for decades many organizations had tried hard to reduce chemical by-products of their operations. The definition includes any solid or hazardous waste that is generated or is subsequently treated, stored, or discarded. The definition envisaged two complementary approaches: reducing the overall volume of the waste and reducing the concentration of toxic constituents in the waste. Toxicity reduction has usually been given priority and is achieved by a variety of means including chemical treatment and incineration. Volume reduction is generally achieved by modifications in manufacturing processes, changes in raw materials, and recycling and reuse. Sometimes reducing the volume of wastes through, say, sludge thickening or dewatering, increases toxicity, but there is less waste to handle. On the other hand, decreasing toxicity through dilution with soil, for example, means there will be more waste to handle.

The real goal of the 1984 legislation was waste "elimination" whenever possible, elimination that must encompass several categories of wastes: raw materials which are not fully used and also the impurities in raw materials; products which are rejected because they are below specifications; useful and useless by-products; and materials that assist in the manufacturing process but have been changed and are no longer useful such as solvents, acids, and catalysts. Clean manufacturing technologies which eliminate all of these problems are the objective of all industrialists as well as environmentalists. However, there are few if any completely clean manufacturing technologies, and capture and reuse of the wastes become more realistic goals.

While the concept of waste minimization evolved from regulation of solid waste, waste reduction programs should obviously be extended to cutting back the generation of gaseous and liquid pollutants as well.

No longer should industry rely primarily on the removal of these pollutants as they pass through smokestacks or discharge pipes to reduce the chemical burden added to the environment. Reduction of pollution should in the first place become an important consideration in designing new manufacturing processes.

Frequently, but not always, the economic savings from waste minimization programs outweigh the costs of introducing the programs. Savings can result from lower costs of handling wastes, reduced requirements for storage areas, lower transportation and disposal costs, and in some cases reductions in state taxes which are levied on the quantities of wastes that are generated. The Chemical Manufacturers Association notes, "Perhaps the greatest long-term economic incentive for waste minimization is to reduce future liabilities and risk. If a waste is not generated, or is generated in smaller quantities, the risk that it might pose to the generator in terms of involvement in a site cleanup or other legal action may be reduced . . . rule-of-thumb estimates place these savings on the order of $100 to $300 per ton."[2]

Turning this very attractive concept of waste minimization into practical engineering approaches takes many forms. Current engineering practices include changes in the characteristics of products or in the manufacturing technologies; recycling, reuse, or reclamation of wastes or their useful components; and reduction through physical, chemical, or biological treatment of the volume and toxicity of wastes that are nevertheless generated.

A leading environmental engineering firm recently cited a few examples of successful waste reduction efforts. With regard to painting in the aeronautics industry, the Hughes Aircraft Company has adopted new dry powder techniques, Lockheed Corporation has substituted many water-based paints for oil-based mixtures, and at Hill Air Force Base innovative techniques for paint stripping dramatically reduce the liquid wastes. As to recycling, a firm in upstate New York has used settling and cartridge filtration to reuse heat-treating quench oil, and an auto assembly plant in the Midwest recycles hydraulic oils using distillation techniques. An electric power utility regularly recovers valuable vanadium from its wastewater, a printed circuit board facility captures copper that previously was being discharged, and at the Charleston Naval Shipyard, chrome is recovered through vapor recompression.[3]

In short, federal and state regulations have increased both the difficulty and expense in disposing of hazardous wastes and have increased the incentives for reducing wastes. Some wastes are banned entirely from land disposal. Other waste disposal methods are tightly controlled, and such limitations have led to a shortage of suitable facilities which are approved for receiving chemical wastes. Of course, operations at these facilities are very expensive, and the costs to the companies which have produced the wastes increase every year.

At the same time, generators of unwanted waste often resist mandatory requirements to change their operational procedures in a prescribed manner. They argue: "The government is in no position to tell us about the economic and technological feasibility of introducing changes in our manufacturing and treatment processes which must be customized to individual facilities. We will make appropriate changes as soon as we can." In many respects these industrial arguments make sense. Still, the government must maintain the pressure on industry to ensure that "appropriate" changes are among the highest corporate priorities.

Looking to the future, as companies calculate costs of raw materials and supplies, they will increasingly include the expense of waste disposal as an up-front cost of conducting their businesses. They will seek substitutes which are less toxic. Sloppy housekeeping practices—leaking tanks, loose valves, faulty pumps, spills, and inadequate cleaning of equipment—will not be tolerated as in the past. The flows of wastewater so often laden with chemical pollutants will undoubtedly be reduced. Recovery of waste products will be encouraged on all fronts. Companies have long been eager to recover waste gold from defective electronic circuit boards, but they have now arrived at the point where they may try to recover sandpaper grit from scrap sandpaper.

Industry Reaches Out to the Public

As industrial plants reduce their wastes and tighten their controls on environmental releases of toxic chemicals, their parent companies are becoming increasingly confident that plant managers can successfully reach out into local communities and improve the public image of the safety of their operations. At some point irate citizen groups, as well as governmental agencies, have besieged almost every major industrial facility for polluting local communities. Steel mills and chemical plants were among the earliest targets of citizen anger. Then the public learned that even the supposedly clean electronics industry with few smokestacks was discarding chemicals into the groundwater. Today we know that the many companies working for the government's nuclear defense industry may be the dirtiest of all. Typically, the response of industry to public accusations of irresponsible pollution and to the associated media blitzes had been simply to comply with regulatory orders issued by the government. When necessary, industrial lawyers argued their cases in the courts. Now industry is clearly in a period of transition as it is being forced into a greater degree of openness.

Of course, the EPA requirement for manufacturing facilities to declare publicly the types and quantities of toxic chemicals they release has been an important stimulant for this change in approach. No doubt many firms believe that the best "defense" against greater public outrage is a good "offense." They now spend time trying to convince local communities that releases of chemicals do not necessarily translate into risks to nearby neighborhoods, that toxic discharges are being cut back, and that there are many economic benefits associated with chemical activities.

A series of interviews carried out by a trade association in 1989 with 20 chemical plant managers in Illinois, Louisiana, New Jersey, Pennsylvania, Texas, and West Virginia has provided a checklist for companies in building their corporate image with a skeptical public. The principal conclusions, together with illustrative comments by plant managers, were as follows:

> Run a safe operation. "If you don't have that, community work is a sham and a fraud."
>
> Reduce the quantity of pollutant releases. ". . . the sheer numbers lead them to draw conclusions about adverse health effects."
>
> Coordinate outreach activities with nearby manufacturing facilities. "Facilities are not viewed separately but as an industry. . . . We educated the smaller plants about the requirements of the law."
>
> Present release information early and share it with employees, government officials, citizens, and the media—in that order. "If you can't convince your own employees, you're not going to convince someone out in the public."
>
> Put your release data into perspective. "We found a local filling station that estimates it emits 20,000 pounds of gasoline vapor a year

as people fill their cars with gas. Using that kind of analogy, we give people a better idea of what the numbers mean."

Open the plant doors. "A group of girl scouts came to the plant at 4 A.M. as part of an overnight activity.''

Be prepared to answer people's questions at all times, and be responsive to people's concerns as soon as they arise; don't wait for a time that's convenient for you. "As people become more aware of their rights to complain, we get more complaints."

Find out the concerns of the community so that you can decide what kinds of community outreach activities to sponsor. "We have approached these problems in the community as if we're dealing with technical problems, when really the problems are ones of perceptions and feelings."

Get involved in the community and "demystify" the chemical industry. "The first thing I always tell my audiences is that we don't run a candy factory."

Contribute time and money to science education in your local schools. "'Chemophobia' is due to people's ignorance of the chemical industry. . . . I ask kids how, using one word, they would describe the chemical industry. They usually say things like 'ugh,' 'cancer,' 'noise,' and 'destructive.' One time a student said 'helpful' and the other kids in this 10th grade advanced science class booed him out of the room."

Embrace your opponents and those who have the potential to become your opponents. "It's more difficult for people to yell at an individual than at those #$%&* across the street.' "[4]

Sometimes, chemical companies try too hard in their public communications to play down the risks associated with their activities. For example, a recent industry brochure likens one part per trillion (presumably referring to dioxin contamination) to a flea on 360 million elephants, a postage stamp in an area the size of Dallas, and one second in 320 centuries. Understanding the smallness of trace quantities of chemicals is important, but using comparisons based on fleas and elephants can only be seen as an attempt to belittle the significance of serious scientific research efforts.[5]

During the next decade, industrial processes in general, and those of the chemical industry in particular, will become even more transparent to the American public. Companies are becoming well aware of the importance of having an informed public and press. Given the historical suspicions of chemical polluters, many companies are working especially hard to contribute to this educational process.

Industry Braces for Transportation Accidents

Every day tens of thousands of tons of chemicals cross America's highways and railroad lines. Chemicals fill pipelines that connect industrial facilities many miles apart. Still larger quantities of chemicals are constantly moving on barges and tankers along rivers and within coastal ports.

Given the volume of chemicals which are always in our transportation systems, spills from transportation accidents are inevitable. Frequently minor collisions of trucks and temporary derailments of tank cars stir the anxieties of police and fire departments. However, sometimes chemicals moving through populated areas are jolted to a point where they explode or burst into flames threatening local residents and passersby. Once in a while, major accidents in harbors, along rail lines or pipelines, or on roads can cause the evacuation of industrial or residential areas or contaminate waterways and drinking water supplies.

Human error will continue to result in accidents and spills that threaten people and ecological resources. A few railroad engineers, ship captains, and truckers will insist on mixing whiskey and work. Some irresponsible shippers will shortcut maintenance procedures on their vehicles and their equipment. Some haulers will simply become lazy and careless.

But to the American public, the inevitability of human failure simply exacerbates perceived dangers of chemicals threatening communities. Each accident reinforces the public's vision of the flammability, ignitability, and toxicity of chemicals. Many chemical manufacturers are now making major financial commitments to railway and trucking companies to support training and inspection programs which can help reduce the incidence of accidents. However, mishaps on the highways, on the railways, and on the waterways will remain an Achilles' heel of the chemical industry in the eyes of the public regardless of preventive measures of individual companies or of penalties imposed for transportation laxity by government agencies.

The reality of highway and railroad accidents and of human shortcomings in transportation was brought home to me on two occasions during my time at the EPA's Las Vegas laboratory.

In early 1982, the EPA's office in Dallas asked our specialists in Las Vegas to immediately provide aerial photography of a train wreck in Louisiana. The laboratory was the national center for using aerial photography to assess environmental problems. This wreck resulted in a series of explosions among 37 railcars, including many carrying highly toxic chemicals. Tank cars had been blown for hundreds of feet, and residential areas within a mile of the tracks were being evacuated. The photos of the accident were needed to help position cranes for clearing away the debris even as the fires continued to smolder. According to press accounts, the train's engineer had been drinking beer with his girlfriend in the engine cabin, and he then passed out leaving the control in her hands. She had simply let the train speed along at about 40 miles per hour over a segment of unstable track which collapsed when the speed should have been reduced to 15 miles per hour.

A short time later, I received an evening telephone call from the California Highway Patrol in Barstow, California, advising me that one of the EPA laboratory's small trucks had overturned. The driver was in the hospital, and unidentifiable chemicals from unmarked containers were leaking onto the highway. We immediately dispatched two experts to the scene, and four hours later they had cleaned up the chemical mess. We, as EPA officials, were embarrassed, to say the least. The accident was unavoidable as the truck skidded on a slippery road, but we had no excuse for not properly labeling even very small shipments of chemicals which posed little risk.

Chemical companies are now well prepared for such accidents. They have for a number of years banded together and mobilized their collective expertise to help minimize the damage once an accident occurs.

Often, when an accident happens, confusion reigns as to the characteristics of the chemicals involved. In some cases, as we have just seen, even the identity of the chemicals is unknown. Is the chemical in a pure or diluted form? Should water be sprayed on the spill? How dangerous are the fumes? Can partially damaged containers be moved? What are appropriate cleanup procedures? While the local fire departments on the scene may have handbooks specifying the properties of the chemicals, firemen are naturally very uncomfortable during their initial encounters with strange substances. Generally, the manufacturer of the chemical knows better than anyone else how it will behave and how it should be handled. Local officials are usually eager to receive authoritative advice.

Therefore, 20 years ago, the principal American chemical producers established CHEMTREC as an information service for those responding to emergencies involving chemicals. A central operator through an "800" number provides specific information on the hazards of more than 560,000 chemicals and trade name products. Drawing on an extensive data bank developed in cooperation with all of the major chemical producers, the operator can advise on what to do and what not to do in case of releases, fire, leaks, or human exposure. The operator also immediately notifies the shipper of the chemical who then assumes responsibility for providing further help. More than 5600 companies and organizations rely on this system as their communications center in the event of an accident.[6]

Often the shipper, working with the producer of the chemical, is not able to send an expert to the scene of the accident to provide advice on safe techniques for capping and patching containers, for transferring the chemical from damaged containers, or for dealing with fires or continuing leakage. The distance may be too great and the time too short. Or the producer may be a very small company without available expertise at the moment it is needed. Therefore, the chemical industry has organized 200 industrial teams and about 50 contractor teams throughout the country whose job it is to respond to chemical emergencies. These teams are generally able to provide authoritative advice on any type of chemical spill. This emergency response network is called CHEMNET.[6]

Of course, federal and state agencies also respond to environmental emergencies, and the activities of both CHEMNET and CHEMTREC are supportive of the governmental efforts. These industrial contributions are very important given the technical complexities which often arise in conjunction with chemical accidents.

The Changed Character of American Industry

In the past, most environmental problems were easily identifiable, and in the words of the EPA's first administrator, "You didn't need a scientific panel to tell you that there was a stench in the air, scum on the water, and garbage on the beach." An abundance of readily available pollution control devices led to discernible progress in reducing air and water pollution. Today the problems of trace levels of toxic pollution are far more subtle and not as easy to correct even though billions of dollars are being spent by industry every year to comply with the ever-tighter regulations to control these chemicals.

For many firms the changes in manufacturing processes that have been necessary for environmental compliance have developed slowly and often implemented with great financial pain. Small firms with limited technical resources and firms with minimal profit margins have had particular difficulties. At the same time, for all companies, the environmental movement has triggered an educational effort to understand the meaning of "risk to the environment" in a chemical age.

Though many political leaders expected that a cooperative coalition between government and industry could be formed to solve pollution problems, they were wrong. The adversarial legal processes that have become the hallmark of today's regulatory systems reign supreme in the environmental field. The EPA and most state environmental agencies emphasize confrontation and not cooperation. Lawyers and lawsuits are the order of the day.

Industry of course has opportunities to participate in the early development of regulations, and at the same time government and not industry should have the final word in deciding the most appropriate approaches to stopping pollution which are then engrained in regulations. However, suggestions of industry are often viewed by governmental officials with suspicion even though the companies may be in a unique position to know which methods of limiting industrial discharges will be the most effective for this end. The EPA and the state environmental agencies need to foster a more cooperative atmosphere. They should give credit to companies that go out of their way to promote environmental protection, and particularly through self-imposed constraints that go beyond the requirements of regulations.

In 1979, about one decade after enactment of the National Environmental Policy Act, *Science* magazine commented on a survey of industrial scientists concerning the rapidly growing responsibilities of the chemical industry to control toxic chemicals as follows: "Attitudes ranged from the view that the chemical industry is in mortal peril to the thought that the present trauma will lead to beneficial results both for society and for the industry. Directors of research contritely admitted past shortcomings in the chemical industry's behavior with respect to its products. They particularly regretted inadequate consideration of the long-term fate of substances and were unhappy about careless errors of some users of chemicals. They now recognize that if misuse leads to untoward effects, they will share the onus."[7]

Just several years earlier, industrialists vehemently argued that the environmental programs of their suppliers and the safety practices of the purchasers of their chemicals were beyond their spheres of interest. Now in the 1990s they fully understand and accept, perhaps reluctantly, the concept that all of industry must be seized with cradle-to-grave control of toxic chemicals. A company's responsibility for chemicals does not begin when the chemicals arrive at the warehouse and does not end once the chemicals leave the shipping dock.

As we look ahead, the concept of "open-ended" liability will probably be the most important driving force in leading industry to cleaner technologies. The onset of diseases from exposure to chemicals may not begin for 10 to 20 years after people come into contact with the chemicals, and companies must be prepared to deal with claims of delayed effects. Chemicals that leave the plant premises may remain intact for decades, and all the parties that handle a chemical as it seeks a final resting place will share responsibility should harmful incidents occur en route. Furthermore, individual chemicals or mixtures of chemicals in and of themselves may be harmless; however, if combined with other factors such as smoking or poor nutrition, they might indeed become serious health risks. Manufacturers and users of chemicals must be sensitive to such subtleties in their handling of chemicals.

Thus, it is not surprising that many companies now operate safety training programs for the suppliers and customers of their products. Some companies have their own hazardous waste disposal sites to be absolutely sure that "their" chemicals will not become problems due to someone else's negligence. A few companies simply will not sell to anyone, regardless of the financial offer, property that has housed manufacturing operations even though the facilities may have been dismantled. These companies do not want to subject themselves to lawsuits by new landlords who in the next century may claim residual effects from negligence by previous owners. Therefore, they often build fences around these vacated premises and pay all property taxes on the vacant lots for the indefinite future.

Finally, with regard to the public's perception of a company's operations, public interest groups began in the 1970s classifying firms according to their environmental records. They even urged divestitures by stockholders from those companies which had particularly bad records.

Most recently, a coalition of environmental, religious, and investment groups developed the "Valdez Principles." These principles, named after the infamous Exxon supertanker which ran aground off the Alaskan coast while under the command of a negligent skipper, are intended to be a code of conduct for companies. The members of the coalition control over $100 billion of pension and other social funds. They have pledged to avoid investments in companies which do not adhere to these principles. Among the provisions of this code are mandatory commitments to include environmentalists on boards of directors, to commission independent environmental audits of corporate behavior, to provide full disclosure of all environmental incidents, and of course to reduce wastes, improve energy conservation practices, and market only safe products and services.[8]

Meanwhile, the Chemical Manufacturers Association, a major trade association for all of the major chemical companies and many of the smaller ones, has developed its own set of general environmental principles to guide corporate behavior. Compliance with the code is a condition of membership in the association. However, these principles do not include requirements for environmentalists on boards of directors or for independent environmental audits. The areas receiving special emphasis include preparedness to respond to spills and accidents, minimization of wastes, and reduction of environmental discharges of pollutants.[9]

In the eyes of the public, industry has always been at the center of the problem of toxic pollution. Now Americans should recognize more fully than ever before that industry must be at the center of the solution to the problem. While government can impose many requirements on industry to reduce discharges of pollutants and to clean up mistakes of the past, chemicals have become such an integral aspect of modern living that industrial initiatives to complement laws and regulations are essential in striking a balance between economic growth and environmental protection. The companies themselves are usually in the best position to recognize many aspects of how their products reach the environment and can cause trouble and to develop techniques to thwart environmental problems in the most economical manner.

We should recall the days of World War II when those industrial plants which made outstanding contributions to the war effort were entitled to fly special pennants awarded by the government over their buildings. Now in the war against toxic contamination, those firms which make extraordinary efforts to ensure the cleanliness and safety of their operations beyond the narrow requirements of the law should be similarly recognized in the eyes of the government and the eyes of the public. A few professional societies and environmental journals present awards to individuals, including industrial employees, who make particularly noteworthy technical contributions to environmental protection. Occasionally, local governments and even state governments single out particularly noteworthy environmental programs of industry. This is an encouraging start. But the federal government and most state agencies remain reluctant to pat industry on the back.

America has come a long way in reversing pollution trends by using a long stick on industry, but the time has come for also holding out a tasty carrot—a carrot of public appreciation for responsible industrial behavior.

8 Exceeding the Absorptive Limits of the Global Commons

As soon as several of the inhabitants of the United States have taken up an opinion or feeling which they wish to promote in the world, they look out for mutual assistance; and as soon as they have found each other, they combine. From that moment they are no longer isolated men.
—Alexis de Tocqueville

Threats to the security of our citizens need to be updated and extended to include environmental degradation . . . we are greening our foreign policy.
—U.S. Secretary of State James Baker

The solution of ecological problems is integral to strengthening international peace and security.
—Soviet Foreign Minister Eduard Shevarnadze

Erosion of the Earth's Protective Shield

I was ready for bed when the doorbell rang at our home in the Washington suburbs on a spring evening in 1977. An elegantly dressed middle-aged lady stood on the steps with a thick document in her hand. "We worked around the clock, and it's ready to be sent to the *Federal Register* tomorrow," she beamed with pride. She was the leader of the network of suburban housewives who provided day-and-night secretarial support for the EPA whenever the staff of the Agency had difficulty preparing long and carefully edited documents in response to short deadlines. She was delivering the final draft of the EPA's proposed regulation to ban the use of chlorofluorocarbons (CFCs) in aerosol sprays, a class of chemicals indicted by scientists several years earlier as causing depletion of the ozone layer of the stratosphere.

The EPA staff was particularly proud of this achievement. Beginning in November 1976, in less than six months they had developed and processed through a reluctant bureaucracy a complicated regulation to deal with a significant environmental threat, albeit of unknown dimensions. To arrive at this point the EPA had spearheaded an interagency task force to limit the use of CFCs under a variety of regulatory authorities. In an unprecedented act of interagency coordination, the EPA (pesticide sprays), the Food and Drug Administration (cosmetic sprays), and the Consumer Product Safety Commission (other sprays used around the house) would each propose in the *Federal Register* on the same day regulations to phase out CFCs used in spray cans.

For the record, CFCs have some very desirable properties. They are nonflammable, noncorrosive, nonexplosive, and low in toxicity. They are stable, soluble, and compatible with many types of materials. They had been used for several decades not only as propellants in aerosol sprays, but also as energy-efficient coolants in refrigerators and air conditioners, as gases that provided the expansion properties in energy-efficient foam insulators, and as important ingredients in solvents for cleaning electrical and mechanical equipment.

In 1976, the annual market value of CFCs produced in the United States was more than $700 million, and the value of goods and services directly dependent on these chemicals was in the range of $10 to $20 billion each year. Approximately one-half of the production was used for aerosol sprays. While the United States was the leader in the manufacture of CFCs with almost one-half of the world's production, other countries were rapidly increasing their shares. CFCs have been and still remain a very important commercial product.

In 1974, two American scientists first predicted in the journal *Nature* that the release into the atmosphere of the widely used CFCs would erode the ozone belt which shields the Earth from excessive ultraviolet radiation generated by the sun. Several scientific committees subsequently confirmed the likelihood of these predictions. Limited measurements of chemical reactions in the upper atmosphere during the mid-1970s began to support the validity of this theory. It seemed likely that on a global basis ozone levels in the stratosphere could decline several percent each decade due to the growing presence of CFCs.[1]

In short, at ground level CFCs are stable and harmless. They generate fine and even sprays, they chill and they insulate, and moreover they are inexpensive to produce. But when they escape into the atmosphere, they slowly rise over a period of several years into the cold stratosphere where they can survive for decades. Then, according to scientists, as they come in contact with water vapor and other chemicals in the low temperature of the stratosphere, particularly in the areas over the north and south poles, they slowly become reactive and sensitive to sunlight. They decompose. The chlorine then attacks the ozone, converting the three molecules of ozone into two molecules of ordinary oxygen.

At the time EPA specialists were preparing the regulation, several government agencies, including the National Science Foundation and the National Aeronautics and Space Administration, were convinced that the theory of ozone depletion was valid although the time scale and the extent of the past erosion of the ozone belt were surrounded by speculation. Medical experts argued that as the ozone levels declined and the intensity of ultraviolet radiation on the Earth's surface rose, the likelihood of skin cancer would increase. But how great would the increase be and what would the increased cancer risk be for any individual? Most of the risk estimates developed by scientists within and outside the government were complicated by reliance on seemingly incomprehensible mathematical equations that simply turned off many policy officials. On the other hand, one understandable estimate of the risk presented by the manufacturers of CFCs was that the effect would be equivalent to moving from the moderate sunshine of Washington, D.C., to the brighter rays of North Carolina.

Agricultural experts were concerned about the possible adverse effect of more intensive ultraviolet light on crops, but they could not provide any estimates of the damage. More worrisome, however, were warnings of some climatologists that a buildup of CFCs could affect the Earth's heat balance and hence alter the global climate. They talked about melting icebergs and changes in the four seasons of the year, but they could only do so in the vaguest terms. Meanwhile, the American public had become aware of the problem, and boycotts of aerosol propellants had begun in several states.

The EPA was facing a dilemma. How could the Agency persuasively defend a regulation that would at the very outset cost American industry hundreds of millions of dollars, a cost which could be estimated with some certainty, when the regulation provided no assurance of significant environmental benefits? We at the EPA had convinced ourselves that even a remote possibility of climatological change from continued use of CFCs warranted regulatory action, but the range of uncertainty was so broad that the Agency would be hard pressed to rest its case on this argument.

Therefore, the EPA decided to steer away from scientific jargon and justify the regulation in broad terms understandable to the public. First, the Agency would emphasize the "possibility" of increased rates of skin cancer. Washington was being bombarded by constant media exaggeration of the tumor-producing potential of environmental chemicals, and prevention of cancer was a powerful political argument in defending environmental regulations.

Second, the EPA would characterize aerosol sprays as "frivolous" and nonessential, arguing that alternative approaches for applications of cosmetics, pesticides, and other consumer and industrial products were readily available. At that time, Madison Avenue had developed a popular line for TV commercials for deodorants, "Get off the can and get on the stick."

The regulation itself would make only very general references to possible harmful consequences of continued ozone depletion besides cancer. The EPA surely did not want to assume the burden of demonstrating future risks in a dialogue with the affected commercial interests, nor was the EPA prepared to quantify and balance risks and benefits.

This strategy for dealing with aerosol sprays used in the United States worked. Eighteen months later, after some initial industry grumbling, the regulatory agencies enacted final regulations for phasing out sprays using CFCs. However, related efforts to establish a more leisurely but firm timetable for addressing the problems of air conditioners, refrigerators, and foam insulation and to mobilize international action to control CFCs even in aerosol sprays foundered in the months and years ahead.

On the domestic scene, by the autumn of 1978 when the limitations on aerosols began to take effect, the American public which had become aware of the problem was losing all enthusiasm—which never was very high—for more expensive substitute coolants for refrigerators and air conditioners, feeling that much of the problem had already been resolved. Hairspray is one thing. Frozen food and air conditioning are something else. Second, American industry had managed to maintain its general level of production of CFCs since the markets for coolants and foam insulators were growing. Thus, the key manufacturers were prepared to go along with limitations on aerosols but strongly resisted further limitations on other products. To add to the inertia, while scientific data continued to accumulate confirming the original theory of ozone depletion, the uncertainties surrounding the severity of the problem remained.

Following the issuance of the proposed regulations in 1977 to phase out aerosol sprays, the United States quickly attempted to take the diplomatic lead. During the summer of that year, the Department of State convened in Washington the first intergovernmental meeting on controlling CFCs which attracted most of the west European governments. American scientists had uncovered a problem affecting all people, and the United States had already taken decisive action involving significant economic costs to begin controlling the problem at home. Many other countries were impressed. However, only Canada, the United Kingdom, Norway, and Sweden took prompt action to limit aerosols while other countries, under pressure from their own industrial constituents, dragged their feet.

In the years that followed, the United States did not press these other countries very hard, since arms control, trade, and international financial arrangements were considered far more important than somewhat vague concerns over the ozone belt. Many Europeans argued that in view of the lack of scientific certainty showing the planet was endangered, they were not prepared to constrain their economies. In any event they were preoccupied with regional air pollution problems in Europe. For a decade the international debates and negotiations were held at low levels with little political clout attached to the discussions among scientists and technicians, and even these discussions became bogged down in haggling among scientists over the uncertainty about atmospheric processes, haggling which the EPA had tried so carefully to avoid at the outset.

Then in 1985 British scientists reported that a hole in the ozone layer over Antarctica had been growing each spring since 1979. The 1984 hole was described as "larger than the United States and taller than Mount Everest." During the next several years, a plethora of scientific reports from American and other scientists using both satellite and ground-based observations issued dire warnings. The hole could continue to spread, and estimates based on the rate the hole was growing suggested the ozone in the Earth's stratosphere would decline by several percent each decade. Scientists argued that this could mean increased cataracts, depressed immune systems, and disruption of sensitive terrestrial and aquatic ecosystems.

Subsequent investigations by other scientists generally confirmed the British findings, and the entire world awoke to the reality of the dangers resulting from an eroding ozone shield. A new sense of urgency led to calls by the Congress and environmental groups for actions in the United States which would limit the use of CFCs as a refrigerant and as a blowing agent in foam insulation. Internationally, by late 1985 the Department of State had elevated the issue of control of CFCs on its list of diplomatic priorities as concerted efforts were initiated by the United States around the globe to reverse the trend in ozone depletion. Initially, American diplomats hesitated to use the "hole" as justification for international action lest subsequent studies rejected the theory, but this hesitation soon disappeared.

In 1986, most of the countries which were the principal manufacturers and consumers of CFCs signed an international convention in Vienna calling in general terms for limitations on CFCs. In the same year, the DuPont Company, the largest manufacturer of CFCs, with one-fourth of the world's production, acknowledged the linkage between CFCs and ozone depletion, and this forthrightness gave considerable impetus to the international effort. Then, after spirited negotiations, in September 1987, 24 countries agreed to an implementing protocol in Montreal calling for a worldwide 50% reduction in CFC emissions by 1998, with initial limitations beginning in 1989. Also, in 1992 the protocol was to freeze emission levels of a class of chemicals called halons. These chemicals are used in fire extinguishers and contain bromine which is even more destructive to ozone than the chlorine found in CFCs.[2]

At a subsequent meeting in London in 1990, the signatories of the Montreal Protocol agreed to more severe restrictions leading to a ban on the use of CFCs and halons beginning in the year 2000. Two other important ozone-depleting chemicals, carbon tetrachloride and methyl chloroform, are also to be phased out by the years 2000 and 2005, respectively. Of considerable importance, China and India indicated a desire to adhere to the Montreal Protocol. They and other developing countries are to have a ten-year grace period to comply with the agreed timetables. In addition, the industrialized countries pledged to contribute to a fund with an initial three-year budget of $160 million to help the developing countries switch to chemicals and technologies which are less damaging to ozone.[3]

The original Montreal Protocol and the subsequent modifications are frequently hailed as a model for promoting international cooperation in environmental issues of global significance. They certainly have been important first steps in paving the way for stringent limitations on global air pollutants, and they have demonstrated the feasibility of reaching agreement on pollution issues that affect the economic interests of many countries.

Scientists are still not sure of the rate of ozone erosion, but the linkage between the presence of CFCs and ozone destruction is unmistakable. Given the current levels of CFCs already in the stratosphere and the emissions that will be released in the years ahead, ozone depletion will continue during the next decade. Had the U.S. government been more forceful in pressing for limitations on CFC production in the United States and in its international negotiations one decade ago, the accumulation of CFCs in the stratosphere probably would have been significantly less. However, now the United States and other countries must move quickly, even if there is only an outside chance that the most dire predictions of studies by agencies of several governments will come true—predictions that the costs to the world in human health and ecological damage from increased ultraviolet radiation during the next century could run into hundreds of billions of dollars.

One particularly encouraging step was an announcement in 1989 made by the DuPont Company that it would terminate production of

CFCs and would market substitute products with greatly reduced impact on the ozone layer. Hydrofluorocarbons and hydrochlorofluorocarbons appear to be the most promising alternatives. They retain many of the desirable properties of CFCs. At the same time, these newly developed compounds either contain no chlorine or decompose before they reach the stratosphere, and their ozone-depletion potential is less than 10% of the potential of CFCs. Still, the revised Montreal Protocol calls for cessation of their use in 50 years, and in the decades ahead we should keep this warning in mind. Unfortunately, early industrial estimates are that these substitutes are less efficient as coolants and may cost as much as five times more than CFCs. However, prices will undoubtedly decline as competitor companies also develop alternative products for the lucrative refrigerant market.[4]

In the meantime, large quantities of CFCs currently found in refrigerators in almost every American home and in air conditioning units in many homes and in three-fourths of our cars will surely escape into the atmosphere. The feasibility of draining such refrigeration systems and inserting new coolants or of containing CFCs when the refrigerators and automobiles are discarded must be explored. However, the outlook for finding efficient ways to capture these chemicals is not bright.

The case of CFCs is a dramatic example of the global reach of man-made chemicals. While useful and benign in our hands on Earth, these chemicals can indirectly affect our lives in many ways once they waft 10 to 20 miles into the sky. Better understanding of atmospheric interactions and of the effects that increasing levels of ultraviolet radiation can have on Earth is imperative. However, governments have recognized that they can no longer hesitate. They must move forward as rapidly as possible to stringently control these chemicals and to begin to check the erosion of the ozone layer.

At the same time, CFCs are only the tip of the chemical iceberg that is slowly drifting into the globe's gaseous shield. While they are currently the principal culprits of ozone depletion in the stratosphere, the Montreal Protocol includes limitations on other chemicals as well. Still other types of emissions from automative vehicles and from industrial and agricultural facilities have also been identified as contributing to the decline of our ozone shield. For example, methane and carbon monoxide are participants in the processes leading to changes in ozone levels. In the words of atmospheric scientists, these gases are among "the precursors to the hydrogen, nitrogen, and chlorine oxides which catalyze the destruction of ozone in the stratosphere."[5] Thus, the U.S. government must think broadly about international action to control an array of environmental pollutants if the ozone blanket is to retain its protective power.

Gaseous Pollutants Warm Up the Earth

Let us turn to the related topic of the greenhouse gases. These gases influence how much of the sun's energy is absorbed on Earth and how much is radiated back into space. The popular conception is that these gases act like a giant greenhouse trapping energy emitted from the Earth below them and causing a warming on the surface of the globe. Unlike the resolution of the CFC problem which was highlighted by the international leadership of the United States, our government is perceived as dragging its heels in international negotiations of limitations on the greenhouse gases as other industrialized nations take the lead.

The greenhouse gases include a variety of man-made air pollutants. Carbon dioxide is the most troublesome greenhouse gas, believed to be responsible for over 50% of the greenhouse effect which can be traced to human activities. Other greenhouse gases are the same pollutants that scientists have linked with depletion of the ozone layer, including CFCs (responsible for an estimated one-fourth of man's contribution to the greenhouse effect), methane (responsible for about 15%), and nitrogen oxides (responsible for a smaller but still significant percentage). On a geographic basis, the principal contributors to the greenhouse effect are, according to one estimate, the United States (21%), the USSR (14%), western Europe (14%), China (7%), eastern Europe (6 percent), Brazil (4%), and India (4%).[6]

When present in the atmosphere, the greenhouse gases allow radiation from the sun to penetrate to the Earth's surface since the molecules of these gases do not interfere with energy at short wavelengths characteristic of sunshine. The Earth absorbs the solar energy, converting it to heat in the process, and then radiates some of that energy at longer wavelengths back toward space. However, the greenhouse gases intercept these longer wavelengths and prevent the escape from the Earth's atmosphere of some of the energy. The Earth currently absorbs about two-thirds of the sun's radiation and radiates one-third back toward space. While a buildup of greenhouse gases in the atmosphere will not alter the percentage that escapes into space very much, even small changes in the dispersion of energy can disrupt the delicate energy balance which sustains life as we know it.

Specifically, the trapping of energy close to the Earth can cause a rise in atmospheric temperatures—global warming. Water vapor and other naturally occurring chemicals which have a greenhouse effect analogous to the effect of the man-made chemicals of concern have always retained energy radiated from the Earth in the atmosphere. This natural trapping has been essential to maintaining the temperatures of the globe above subfreezing levels. However, with the increasing buildup of greenhouse gases due to industrial and agricultural activity, temperatures are now slowly rising to higher levels. The rate at which temperatures are increasing or may increase in the future as the gaseous pollutants accumulate is of course a central question.

As we have seen, increased ultraviolet radiation reaching the Earth after the loss of stratospheric ozone due to CFCs has caused great concern in many corners of the world. In recent years, predictions of the consequences of global warming have sounded an even more ominous chord. With each new measurement of increased levels of carbon dioxide in the atmosphere, anxieties heighten. During the late 1980s, the hot summers and the droughts in the United States and elsewhere led many political leaders to believe that the greenhouse effect was already upon us and that real estate investments would soon shift northward. Some scientists, however, argued that the temperature and moisture excursions of the 1980s were simply fluctuations in the perennial cycles of droughts and floods.

If we look back in time, we can see that the concentrations of carbon dioxide and other greenhouse gases in the Earth's atmosphere have increased steadily during the past 100 years—a period of intensive industrialization in many countries. Still, reviews of temperature records during that time have led to conflicting conclusions by the experts as to whether there have been "significant" temperature changes which correlate with the increasing levels of carbon dioxide in the atmosphere. Nevertheless, according to some scientists, by the middle of the next century levels of the greenhouse gases could be twice current levels, and temperatures could rise at least several degrees Fahrenheit. To provide a perspective, we should note that the temperatures of the ice age 10,000 years ago were about nine degrees lower than at present.

Most scientists are reasonably certain that they have identified the most important chemical contributors to rising temperatures. However, scientific understanding of the processes that lead to global warming is far from adequate. These chemical reactions are complicated and interrelated. Similarly, projections of the consequences of global warming are fraught with uncertainty.[7]

Specialists have constructed elaborate computer models to simulate the conditions on the Earth. Several models have predicted that as a result of the buildup of carbon dioxide and other gases, the temperature of the planet may increase by at least several degrees. If modest changes in the global temperature of four or five degrees occur over several centuries, societies around the globe might be able to adapt to the new environments without major disruptions. However, if such temperature changes occur over a period of only a few decades, societies could have great difficulty adjusting to the new conditions. According to the models, the level of the seas could rise several feet. Some coastal areas would be flooded—in Florida, Long Island, and Louisiana and in low-lying countries around the world such as Bangladesh, Egypt, and Vietnam. As ocean currents change course, storm patterns shift dramatically. Also, temperature and precipitation changes can disrupt agricultural practices. Based on the 1988 experiences, American agriculture would be particularly hard hit. Also, as an extreme, populations in some areas of the world might be forced to move into regions where unfamiliar diseases can attack those lacking natural immunities.

However, if scientists cannot agree on the extent that air pollution levels have influenced surface temperatures in the past, can they realistically predict such linkages in the future? The uncertainties as to the precise relationship between the buildup of greenhouse gases and temperature changes are enormous. Given the stakes involved, everyone agrees that greatly expanded research efforts to better understand the processes affecting the character of the atmosphere are urgently needed. Indeed, almost every governmental research agency throughout the world now has a "global change" research program.

One of the most authoritative statements of scientific understanding of global warming is the 1990 report of an international panel of governmental experts assembled under the auspices of the United Nations Environment Program and the World Meteorological Organization which concluded that:

> We are certain emissions resulting from human activities are substantially increasing the atmospheric concentrations of greenhouse gases: carbon dioxide, methane, chlorofluorocarbons (CFCs), and nitrous oxide. These increases will enhance the greenhouse effect, resulting on average in additional warming of the Earth's surface. The longer that emissions continue at present day rates, the greater reductions would have to be for concentrations to stabilize at a given level. The long-lived gases would require immediate reductions in emissions from human activities of over 60 percent to stabilize their concentrations at today's levels.
>
> Based on current model results, we predict under the Business-as-Usual (i.e., no change in current practices) emissions of greenhouse gases, a rate of increase of global mean temperature during the next century of about 0.3 degrees centigrade (or 0.5 degrees Fahrenheit) per decade with an uncertainty range from 0.2 to 0.5 degrees centigrade (or 0.3 to 0.9 degrees Fahrenheit) per decade, greater than we have seen over the past 10,000 years; under the same scenario, we also predict an average rate of global mean sea level rise of about 6 centimeters (or 2.4 inches) per decade over the next century with an uncertainty range of 3 to 10 centimeters (or 1.2 to 3.9 inches) per decade. There are many uncertainties in our predictions particularly with regard to the timing, magnitude, and regional patterns of climate change.

Thus, these experts predict that unless the emissions of greenhouse gases are significantly reduced, within 50 years the Earth's average temperature will increase by 2.7 degrees Fahrenheit, although it could increase by as much as 4.5 degrees; and the level of the seas will rise by about one foot, but they could rise by as much as 1.6 feet. The experts also concluded that:

> Rapid changes in climate will change the composition of ecosystems; some species will benefit while others will be unable to migrate or adapt fast enough and may become extinct . . . . In many cases, the impacts will be felt most severely in regions already under stress, mainly developing countries. The most vulnerable human settlements are those especially exposed to natural hazards, e.g., coastal

or river flooding, severe drought, landslides, severe storms, and tropical cyclones.[8]

In addition to studies of the problem, what should be done now to *control* the greenhouse gases? Can nations simply delay corrective actions to reduce emissions of the greenhouse gases while waiting for conclusive scientific evidence as to the magnitude and urgency of the problem—evidence that may never be adequately developed or developed too late to alter irreversible trends?

During the initial years of the Bush Administration, tempers flared as U.S. government officials debated the alternatives. The White House staff repeatedly rejected proposals of the EPA and the Congress for the United States to join with other nations of the world in establishing a quantitative worldwide goal for reducing emissions of carbon dioxide. They were concerned over the economic costs to U.S. industry and to consumers from compliance with such a commitment. Initially, the United States, together with the USSR and Japan, became isolated from the other countries by refusing to make the political commitment to such aggressive regulatory action for attacking the greenhouse problem. While these three countries argued that they would take steps to limit emissions, their refusal to endorse a specific target was widely perceived as a political message that the world's leading polluters would continue to consider economic growth more important than the global ecology. Then in 1990 other industrialized countries, undoubtedly influenced to some degree by the position of the United States, also became less enthusiastic about committing to sharp reductions of carbon dioxide.

It seems clear that global temperatures are rising due to human activity and that if unchecked, the increases will eventually cause global disruptions—perhaps in 50 years, perhaps in 100 years, but sooner or later. A variety of approaches are available to reduce the buildup of carbon dioxide and other greenhouse gases, in addition to the important step of phasing out CFCs. They include reducing leakages from pipelines and other man-made sources of methane, shifting from "dirty" coal to "clean" gas, and adopting a variety of energy conservation measures. There is a growing consensus to follow such approaches which will clearly be helpful, but they may not be adequate for slowing the current accumulation of greenhouse gases.

The Bush Administration supports restraints on air pollution that can be justified on the basis of reducing energy costs or reducing health risks. Such steps can at the same time help abate the greenhouse effect. This is called the "no regrets" policy since its success in terms of environmental improvement is not dependent on the uncertain outcome of scientific deliberations over the extent of global warming. The administration reinforces this policy of going slow with regulatory actions by calling for annual research investments of $1 billion (eventually reaching $3 billion) to improve the scientific base for predictions of the likely impact of pollutants on the long-term climate.

Still, the concept of quantitative reduction targets for emissions of greenhouse gases must be faced squarely by the United States, as well as other nations. The concept will not go away. For example, beginning in 1989, several European governments (e.g., Sweden, Norway, West Germany, England, and Denmark) committed themselves to targeted reductions of carbon dioxide. They recognized that their individual actions would be insufficient, but they hoped to be symbols which would inspire others to act. Will the international community, including the United States, make a serious political commitment to reducing greenhouse gases, a commitment that will be believable to many only if it is couched in terms of meaningful emission reduction targets? Living up to such a commitment may be expensive, and nations must be prepared to slow down economic prosperity, at least as we currently define prosperity, today to ensure the well-being of future generations? In 1992 the United Nations will hold a global conference in Brazil on environmental and development issues, and the United States should be prepared to adopt specific reduction targets as well as endorsing the continuing need for research.

In short, society cannot rely on the chance that the 1990 predictions of a likely rise in global temperatures were exaggerated and that the experts underestimated the absorptive capacity of the planet. The longer the delay of corrective actions, the higher are the costs of these actions. More importantly, some harmful effects during the period of inaction may be irreversible and not correctable. Expenditures now to curb carbon dioxide emissions are an insurance policy we can well afford. A "no regrets" policy is a first step, but a timid step that should be supplemented with an insurance policy of additional actions just in case some of the dire predictions of impending disasters turn out to be warranted.

Coping with Fossil Fuels

Because carbon dioxide emissions from the combustion of fossil fuels play a central role in the greenhouse debate, policy officials throughout the world have focused their attention on ways to reduce carbon dioxide discharges associated with energy production. Most of the current emissions from fossil fuels are attributable to the United States and other industrialized countries. California alone acknowledges responsibility for discharging 3% of the greenhouse gases worldwide, including large quantities of carbon dioxide from fossil fuel combustion. Unfortunately, prospects have not been bright for reducing carbon dioxide releases during the next 50 years in a world bent on greater economic affluence and driven by energy-intensive technologies. Still the power of financial incentives should not be underestimated. At the time of the high OPEC oil prices in the 1970s, industrial organizations throughout the world adopted many measures to reduce their needs for energy supplies and clearly demonstrated that when motivated, they can reduce wasteful consumption patterns of the past.

Popular magazines and the press throughout the world have been inundated with articles depicting a planet suffocating under a cloud of power plant emissions. As a result of both the findings of scientists and this media blitz, reducing the demand for energy and introducing cleaner technologies for generating energy are now attracting long-deserved political support in the United States and elsewhere. At the federal and state levels, officials are calling for larger budgets and stronger laws to provide the technical and legal basis for eliminating wasteful energy practices. Let us hope this support will not fade as it did in the late 1970s when the OPEC oil prices declined and Americans became overly optimistic that new versions of the Clean Air Act would lead to adequate reductions of emissions of many harmful pollutants.

To most Americans, cutting back the demand for energy means driving smaller cars, erecting buildings which have better insulation, adjusting heating and cooling practices in homes and other structures, and increasing efficiency in electrical power plants and in industrial facilities. However, despite growing political interest and public awareness of the importance of energy conservation, the gap between actual practice and the opportunities for conserving energy is large. The public has taken many easy steps to reduce energy usage and cut utility bills. Nevertheless, large cars, high ceilings, glass buildings, and bright lights seem more popular than ever. Meanwhile, some American industrialists argue that due to economic considerations many inefficient plants cannot be replaced before the end of their normal lifetimes of 30 or more years and that in the interim they have no choice but to operate facilities which were not designed to minimize energy consumption requirements.

In the immediate future, common sense and a modest amount of political will are all that are needed for the United States to introduce more aggressive policies toward energy conservation. Unfortunately, voluntary actions on the part of the American public and our industry in response to general policy pronouncements to save energy have not been adequate in the past. A combination of regulatory and economic incentives will undoubtedly be required. Increased taxes on energy supplies, expanded governmental support for research on energy-saving technologies, and regulatory pressures to reduce fuel consumption of automobiles and trucks are steps that the government can take to encourage much improved energy performance of the nation's fleet of motor vehicles and more effective conservation practices of homeowners and businesses.

As discussed earlier, the government must enforce more stringent requirements for controlling discharges of air pollutants from industrial facilities. This will limit human exposures to harmful chemicals at ground level as well as help mitigate the greenhouse effect. These requirements should, in addition, encourage greater attention to reducing energy wastage when designing and operating industrial facilities.

With regard to cleaner energy technologies, the wind and solar energy euphoria of the 1970s has given way to the harsh economic realities of the 1980s. These technologies simply are not likely to become major contributors to energy generation in the foreseeable future. Still, the Washington-based Worldwatch Institute continues to predict that "direct conversion of solar energy will be the cornerstone of a sustainable world energy system."[9] If such an approach could become technically and economically feasible, everyone would surely applaud. No other source can rival the sun as a clean and inexhaustible supply of energy. But few experts share the institute's optimism over the likelihood that photovoltaic cells, solar reflectors, and wind farms will be the anchors of electrical grids that support the megacities of the next century. There are simply enormous technical problems in collecting sunlight from broad areas, concentrating it, and distributing it to consumers in large quantities.

Looking ahead 20 years, the energy mix in the United States is most likely to be of the following character: petroleum, 41%; gas, 20%; coal, 25%; nuclear, 6%; hydro, 3%; and solar, biomass, and other renewables, 5%.[10]

In short, a variety of political, economic, and technical measures must be taken to promote energy conservation and to discourage the use of dirty energy technologies during the 1990s. In the longer run, the United States should seek more dramatic changes in its approaches to the generation and use of energy. Unfortunately, one important alternative to fossil fuels, more nuclear power stations, remains in limbo in the wake of the accidents at Three Mile Island and Chernobyl.

Reviving the Nuclear Option

Nuclear power stations offer an attractive alternative for clean energy, devoid of greenhouse gases. However, debates over the lifetime expenses of constructing, operating, and dismantling nuclear facilities, as well as sharply differing views on nuclear safety, divide the advocates who claim large savings from use of nuclear power from opponents who are convinced that the technology is dangerous and the costs are out of control. But as clearly demonstrated in France where nuclear reactors provide two-thirds of the nation's electricity, the technology long ago advanced to the point where the costs and safety are sufficiently competitive with fossil fuel plants to warrant serious consideration of greater reliance on nuclear power.

As discussed before, the technical problems associated with nuclear wastes are manageable if our nation adopts a strategy for the next century of storing contaminated nuclear fuel rods in lead casings on the

Nevada desert rather than now pressing for their permanent disposal in deep geological caverns. However, the populace in many countries remains skeptical over safe operation of nuclear plants. Several personal experiences reflect the basis for this skepticism.

In 1958 while serving at the American Embassy in Belgrade, Yugoslavia, I helped mobilize medical assistance to the victims of an accident at a nuclear research reactor at the edge of the city. The nuclear engineers had decided to operate the reactor even though the automatic control system was being repaired. They failed to properly handle the manual control system, and several deaths ensued when excessive radiation was released despite the valiant efforts of French doctors to save the victims through bone marrow transplants in Paris.

Then, in 1961, while working in the nuclear engineering division of Argonne National Laboratory near Chicago, I took a trip to see firsthand the remnants of an explosion in a nuclear reactor built by the U.S. Army in a remote area of Idaho. The press had reported some of the design and operational flaws leading up to this tragedy which killed several workers. Rumors circulated that one of the reactor operators deliberately caused the accident in a jealous rage over an affair between his wife and another operator. Meanwhile, the national security blanket which was placed over both the reactor and the accident raised many suspicions that safety had not been given a high priority.

At the EPA in 1981 I became immersed in the aftermath of the accident at Three Mile Island. Several reviews of the events highlighted the unlikely scenario of technical mishaps that triggered the reactor's failure. Mechanical devices simply did not work properly, and the reactor's operating crew was not prepared to cope with unexpected technical failures. While there were no physiological effects on the nearby population from the accident, the psychological impacts were serious and linger to this day.

Finally, during several trips to the USSR in the late 1980s, I observed how an aroused public, irritated by the cavalier operating procedures which led to the explosion in the Chernobyl reactor, was trying to dismantle the Soviet nuclear industry. At Chernobyl, more than 30 workers and rescue personnel died, tens of thousands of inhabitants of the region were exposed to radiation levels which will probably increase the incidence of cancer and other diseases among this population, and efforts to clean up large contaminated areas many miles from the site will continue for decades.

This track record is attributable in large measure to human failures but even more fundamentally to flawed designs of technological systems which did not give adequate weight to the possibility of such failures. These and other accidents understandably raise many apprehensions around the world about the desirability of continuing, let alone expanding, reliance on nuclear power. Indeed, in recent years many plans for nuclear power plants have been scrapped at home and abroad. Only a few new plants are now coming on-line in Korea, France, and Czechoslovakia, for example.

However, abandoning the nuclear energy option at a time of growing concerns over the greenhouse gases let alone uncertainty over the reliability of petroleum supplies from an unstable Middle East makes little sense. Few advocate closing chemical plants because of the Bhopal tragedy and many other industrial accidents, ceasing ocean shipments of petroleum products because supertankers occasionally run aground, or outlawing explosives which are sometimes misused. The stakes to society in the future of nuclear power are of a similar magnitude of importance. Improved safety features rather than efforts to dismantle the industry should be the order of the day. Such safety must emphasize an extraordinary degree of attention to the training of nuclear operators given the sensitivity of current nuclear systems to human failure.

Nuclear materials can be very hazardous and require special precautions. In addition, the crossovers between military and civilian nuclear activities have long called for measures to prevent diversion of nuclear materials from civilian to military applications, particularly in countries with terrorist tendencies. Also, the lengthy periods that many potent radionuclides survive in the environment underscore the great care that is needed in following these materials to their ultimate disposal site.

During the 1980s the safety procedures within the United States for designing and operating nuclear power stations were overhauled. Minor discrepancies in accepted engineering practices which might have been ignored in the past are now considered major violations of operating procedures. The qualifications and training of reactor operators have been greatly strengthened.

In mid-1989, I observed one of the hundreds of training responses to simulated reactor failures staged each year at the Millstone Nuclear Reactor Complex in Connecticut, as well as at other nuclear plants. These tests provided impressive evidence of the soundness of today's approaches for honing the skills of operators that the U.S. government has adopted. In short, American industry has come a long way in learning how to construct, maintain, and operate nuclear plants safely. The United States needs to diffuse this experience widely to those countries which rely on nuclear power but still have weak safety programs.

Meanwhile, the feasibility of building nuclear power plants which are technologically immune from human failures that could lead to catastrophies seems near at hand. A new generation of "inherently safe" reactors is dominating the drawing boards at many engineering installations in the United States and abroad. The design concepts are based on a simple principle: when a reactor begins to heat up unexpectedly, it shuts down independently of any actions taken by the operator. Moreover, the costs of this refinement of nuclear power reactors are projected to be competitive with fossil fuel costs. Our government should encourage the rapid development and testing of such inherently safe systems.

Of course the sociological impediments to a new generation of nuclear reactors will be severe, even if the environmental advantages seem clear. However, the increasing reality of greenhouse warming is beginning to modify some of the public's negative attitudes toward nuclear power. For example, the National Audubon Society now apparently supports the idea of seriously exploring inherently safe reactors. Other important public interest groups are also beginning to recognize the advantages as well as the liabilities of nuclear power. Assuming that these new systems live up to their advanced billings, a reasonable target would be to double the power generation capacity of nuclear stations throughout the world by the year 2020.

At the same time, governments should not become so enamored of these new approaches that they lose sight of the potential problems with the several hundred reactors on-line in many countries that do not incorporate such "fail-safe" design features. Some of these early reactors will remain in operation for several decades, and attentiveness to their operations is imperative. Unfortunately, the technical skills and legal requirements for operating a nuclear industry in some countries are not as well developed as in the United States. Therefore, additional reactor accidents around the world seem inevitable during the 1990s. Hopefully, they will be contained with minimal environmental leakages. Another "Chernobyl" could indeed be the political death knoll for nuclear power regardless of the alternative threats posed by the greenhouse gases.

The current policies of the U.S. government toward the continued development of nuclear power as an important energy source during the next century are sound. Strong support for the development of the next generation of reactors together with stringent requirements on the operation of existing reactors makes considerable sense. At the same time, efforts should be expanded to reduce the costs of solar power and to search for other potential breakthrough technologies.

Too Many People around the World

The energy consumption trends in developing countries are taking on increasing significance. With unchecked birthrates in many countries, the global population is already approaching six billion people, and another billion will be added during the next decade. As some of the poor but heavily populated countries slowly industrialize, the energy demand per person increases as well. In particular, China has large coal reserves. All indications show that coal consumption in that country will double by the end of the century as the population continues to grow despite stringent attempts of the Chinese government to limit the size of families to one child. Similarly in India, consumption of coal is increasing and will likely triple in the next twenty years. Many other developing countries, while less well endowed with fossil fuels, are no less determined to provide electricity for the countryside and are seeking energy supplies for this effort. The implications for global warming are obvious.

Most American energy specialists are convinced that population growth and its attendant energy demands must be curbed in the Third World. However, the U.S. government is hobbled in its programs for developing countries by controversies in the United States over the extent of our commitment to foreign aid in general and family planning in particular and over the most effective and appropriate approaches to interacting with developing countries.

Americans have great difficulty recognizing that their self-interests are increasingly entwined with the fate of billions of people in the Third World. Frequent public opinion polls reveal that many Americans ask, "Why should we help the poor in other countries when we can't take care of the homeless in the United States?" Even our political leaders do not appreciate the extent that money spent abroad now can reap benefits at home in the future, for they repeatedly place unrealistic budget constraints on foreign aid activities. As far as the style of U.S. relations with the poorer countries is concerned, our governmental agencies need to realize that the leaders of these developing countries are determined to shape their own futures. True *partnership* must replace American *patronage* as the key to effective American programs in Asia, Africa, and Latin America.

Even with realigned and strengthened policies and programs toward the Third World, the United States would be able to assume a credible leadership role in seeking reductions of greenhouse gases only when the dividends from more aggressive pollution reduction policies at home become apparent. As long as the United States continues to lead the world in discharges of greenhouse gases with no downturn in current trends, American diplomats will have great difficulty persuading Brazil and other countries to slow population growth and to stop ravaging the tropical forests which absorb carbon dioxide.

Thus, the United States should set an example in energy conservation and pollution control for others to follow. We must also share American scientific and technical skills with the poor countries to provide needed tools for reversing the continued growth of populations and to strengthen worldwide capabilities for conserving forests and other renewable resources of the tropics.

Until the late 1970s the American foreign aid program was a pacesetter in the field of family planning in the Third World. American representatives spoke out repeatedly and eloquently on the advantages of constrained population growth. Then family planning programs became ensnarled in our domestic politics over the propriety of different approaches—particularly abortions.

The complexities of economic development and population growth in undeveloped countries demand attention from the best pool of scientific talent that the industrialized countries can offer. The motivations underlying the continued rearing of large families in poor countries are manyfold and will be difficult to change. Yet there is no alternative to change. The United States need not become involved in controversial abortion programs which are not supported by the American people. Reducing infant mortality, for example, is an objective everyone can endorse, and we have much to offer developing countries in this area. In the long run, improving the chances of children outliving their parents and thereby being available to care for the elderly may reduce the pressures on parents for many sons and daughters to ensure the survivability of one caring offspring. Also, programs to enhance the education and status of women in the poor countries can over time become important deterrents to excessively large families.

Given the enormity and possible effects of population growth, the United States should greatly increase its financial and technical commitment to help developing countries achieve a balance of population growth and economic and environmental sustainability. The United States should no longer be at the bottom of the list of industrialized countries which contribute to foreign economic assistance (measured in terms of percentage of GNP devoted to foreign aid). Americans have much to gain from more rational development of the Third World.

American resources, together with the resources of other industrialized countries and the developing countries themselves, should be focused on specific goals such as halving world population growth by the end of the century. In the absence of such a clear objective supported by many nations, there is little likelihood that the rate of global population growth will slow significantly. While the Bush Administration and the Congress may take solace in knowing that the United States contributes 40% of all international funds directed to population programs, they should recognize that the total funds from all countries are so small that the likely impact on population growth rates during the 1990s will be marginal at best. At the same time, more funds from the United States, say on the order of $500 million per year, together with a together with a more outspoken and aggressive American leadership role in urging all countries to support family planning, could make a significant difference.

In short, global warming is bringing a new dimension to public policy debates in the United States and around the globe. These debates affect almost every segment of society as we seek environmentally safe technologies. As predictions of global warming gain greater credibility, the interest in energy conservation, nuclear power, and limitations on population growth will continue to increase. All of these approaches deserve special attention in our efforts to curb the erosion of nature.

The Growing Role of Eco-Diplomacy

Depletion of ozone in the stratosphere and greenhouse warming are examples of the indirect reactions man-made chemicals can trigger when they rise to high altitudes. These reactions have profound long-term implications for the ecology of this planet and for human survival.

A third well-known example of an air pollutant with international repercussions is acid rain. Acid rain is traced to sulfur oxides which are emitted by power plants, automobiles, and some industries. Sulfur oxides lead to atmospheric transformations of chemical structures that can result in severe damage to forests, lakes, and buildings, often hundreds of miles from the sources of the emissions. For more than a decade, acid deposition in Canada which is attributable in part to activities in the midwestern United States has been a dominant topic in U.S.–Canadian diplomatic exchanges. In the closely compacted countries of Northern and Central Europe, forest destruction linked to acidification has reached alarming levels with acrimonious statements of one nation regularly accusing others of intolerable industrial practices.

These three problems caused by common air pollutants that know no boundaries increasingly dominate the agendas of important international gatherings. A sample would include the economic summit of several Western heads of state in Paris in 1989; a gathering of some of the world's political leaders convened by Prime Minister Thatcher shortly thereafter; the convocation of scientists and religious leaders called by the Pope the same year; an international environmental extravaganza hosted by Mikhail Gorbachev in Moscow in early 1990; a White House conference on global ecology several months later; and the Western economic summit in Houston in July 1990. At all of these meetings participants pledged ever greater efforts in the common struggle against mankind's own transgressions. The cold war pressures for military buildups are rapidly giving way to the pressures for ecological accommodation.

Returning to the more immediate effects on humans and ecological resources of toxic chemicals, the international community, and particularly the Western countries, have for many years been concerned with chemical contamination of foods that are shipped across international boundaries. Occasionally, pesticide residues have been a particularly serious problem, and imported agricultural products have been rejected at U.S. ports. Fortunately, for several decades an international scientific forum called "codex alimentarius" has developed guidelines on acceptable trace levels of chemical contaminants in food and on appropriate methods for measuring such contaminants. These guidelines, which are now accepted by most governments, have been very important in ensuring that chemical poisoning has not become a significant impediment to agricultural trade.

As to industrial chemicals, Western nations were alerted to new types of environmental problems during the 1960s when large Japanese populations suffered food poisoning from cadmium, mercury, and PCBs. Though these incidents were localized problems, they foreshadowed problems with the same chemicals in the United States and Europe. Specifically, in Japan food became contaminated from industrial discharges into fishing areas. In Japan and in other countries encountering similar problems, national actions could be taken without the necessity of concerted international efforts. Nevertheless, the similarity of problems in industrialized areas of many countries and the commonality of the remedial actions developed by individual governments were striking. Most countries facing these problems, including the United States, were eager to pool their knowledge so each could learn from the other.

Meanwhile, European countries were banding together to control pollution in the air and rivers which crossed international borders. In the 1960s European environmental concerns focused on very specific geographic regions which were defined by the mobility of the pollutants. The Rhine River and the Ruhr Basin, for example, became identifiable pollution zones. Even though water basins and air sheds cross international boundaries, the responsible parties and the affected parties were easily identified. The immediately concerned governments worked out solutions, so they thought. Gradually, however, the difficulty of tracing sources and effects in the long-range transport of pollutants in their original or modified forms was recognized. Now regional pollution problems that have subtle impacts on many countries frequently dominate European environmental debates.

Beginning in the 1970s, Japan, the United States, and the European countries enacted laws to control the manufacture and distribution of a few important chemicals which had highly toxic properties. In the United States, this law was the Toxic Substances Control Act. One aspect of these laws requires an exporting nation to inform potential importers of the presence of chemicals in international shipments and of the properties of the chemicals. These national laws also provide authority for rejecting chemical imports for environmental reasons if necessary.

The economic implications of such laws were obvious from the outset since worldwide trade in bulk chemicals exceeded $100 billion in the 1970s and has grown steadily ever since. While statistics have not been available on worldwide trade in items which are the end products of chemical processing industries, in a sense all goods are chemical products, and chemical decomposition and chemical leakages of end products as well as bulk chemicals occasionally occur during shipments. Thus, the scope for regulation of international trade is very broad.

American officials always worry about international actions which could erode U.S. technological leadership that undergirds much of our export strength. In particular, new agreements requiring disclosures by American firms to other nations of proprietary information on the chemical composition and related properties of many substances and products could affect our competitive edge in the chemical field. Also, if laws regulating research and development in the United States or abroad became so stringent as to dampen innovation within the chemical industry, an area where our national and multinational firms excel, the United States could become a big loser on this front.

During the past 15 years the United States has aggressively sought an international consensus on the character of laws and regulations to control the development, manufacture, and distribution of chemicals—a consensus which will not adversely affect American interests. While representing the United States at discussions of such issues at the

Organization for Economic Cooperation and Development in Paris during the early formulation of these laws in the 1970s, my EPA colleagues and I were continuously confronted with the requests of the West European countries for agreement on more restrictions on the development of new chemicals—an area of great U.S. strength. The same countries opposed efforts to review the possible hazards associated with chemicals which were already on the market. These chemicals had found their niches in Europe, and the European governments were not eager to sacrifice them. While representatives of all countries were presumably motivated by environmental concerns, priorities were surely influenced by economic interests. Only now have the European governments finally accepted the concept of phasing out well-established chemicals which are particularly hazardous to health or ecological resources, but serious *actions* in this area are yet to come.

Discussions concerning toxic substances with our European trading partners have concentrated on several areas. Are the tests conducted in the laboratories of one country, say a European country, acceptable to another country, the United States for example, which is determining whether the chemical can be sold in the United States? More specifically, do scientists of both countries agree that the correct tests are being performed and that the testing procedures are of high quality? Second, can delays in imports and exports of chemicals due to new international trade requirements be minimized? Will the customs services throughout the world help ensure that chemicals don't sit indefinitely in warehouses due to the lack of proper documentation or trained personnel for facilitating their movement? Finally, will differing restrictions in different countries encourage multinational chemical companies to locate their test marketing activities in those countries with the least stringent regulatory requirements? What are the economic and environmental consequences of such practices?[11]

In general, the European nations participating in these discussions in Paris recognize the importance and relative ease in sharing scientific expertise and results of scientific research relevant to the control of toxic chemicals. However, discussions of economic and trade issues are more constrained since proprietary information is frequently at the heart of meaningful dialogues. Thus, scientific rather than economic themes dominate. Science is a critical factor underlying regulatory approaches, and a thorough airing of the less controversial issues of science and scientific methodologies will continue to be important. At the same time, reaching consensus on the more difficult regulatory issues is also obviously important and needs even greater attention in the future.

In 1992, 12 West European countries with a population exceeding that of the United States will "integrate" their economic policies. Their consolidated approaches to international trade will have numerous repercussions. The United States should expect further efforts among the Europeans themselves for common approaches to regulation of toxic chemicals. Not surprisingly, many American firms are expanding their commercial alliances with West European partners. These alliances will provide the firms with flexibility in locating manufacturing and research facilities in Europe when appropriate. Also, such linkages should ease shipment of chemicals and other products across international boundaries, and more generally put companies in a position to take advantage of those national regulations which are least disruptive to their normal business practices.

Environmental Negotiations at All Levels

The Organization for Economic Cooperation and Development is just one of many international organizations giving high priority to man-made chemicals that find environmental pathways across national boundaries. More than one dozen agencies of the United Nations, another dozen international scientific organizations, and still another dozen well-established international research centers bring together thousands of scientists and policy officials from around the world every year to consider many aspects of chemical pollution. Environmental issues command an increasingly prominent position on the agendas of international business organizations. Some international financial institutions such as the World Bank have made major commitments to linking their industrial loan activities with the environmental sensitivity of loan recipients. Others give increasing weight to the feasibility of environmental problems in determining the credit ratings of governments, with the USSR and Eastern Europe being recent losers in trying to retain high ratings.

Meanwhile, many prominent public figures from Hollywood to

Wall Street have taken up the cause of worldwide environmental protection and derive great pleasure in using their high visibility for the cause. For example, Robert Redford hosts meetings of international leaders in resource conservation and global warming at his highly photogenic retreat in Sundance, Utah. He has surrounded himself with many environmental experts and has made many useful contributions to the international dialogue on environmental issues.

Environmental diplomacy has entered the foreign policy arena in other ways. For decades, environmental protection of the Great Lakes has been a recurrent theme in the diplomatic debates between the United States and Canada. Many successes have been recorded in limiting discharges into the lakes of sewage, detergents, and asbestos, for example. In recent years, great attention has been directed to PCB contamination which still plagues the lakes. To the south, our Department of State remains frustrated in its efforts to limit waste discharges from Mexico which pollute the beaches and streams of southern California. Also, air pollution from Mexican industries along the border are frequently a bone of contention. In Europe, many countries have signed agreements to limit chemical discharges into the Mediterranean and Baltic seas. While these international commitments have undoubtedly tempered some temptations to avoid pollution controls, many beaches remain closed. All of these efforts are commendable, but we need to realize that environment commitments do not always equate to environmental improvement or even to compliance with such commitments.

Bilateral agreements between the United States and many countries to promote cooperation in environmental research have become very fashionable. For example, the East–West political dialogue increasingly includes discussions of ecological issues. The United States and the USSR, for example, now have more than 40 formally negotiated cooperative projects in the environmental field, as well as many more less formal arrangements. As to Eastern European, the recent political reforms were stimulated in part by citizen anger over environmental deterioration, and now unprecedented opportunities exist for international cooperation and assistance in one of the world's most highly polluted areas.

As noted, a summit of environmental leaders from many countries is scheduled in Brazil in 1992 under the auspices of the United Nations.

Much of the current international effort is directed to preparing treaties which can be signed in connection with that meeting. At the top of the list is a treaty to put a quantitative cap on discharges of greenhouse gases. In the interim, several steps have already been taken to cut back air pollutants on an international scale. In particular, agreements have been signed by a number of countries to reduce discharges of sulfur oxides and nitrous oxides. However, gaining wider acceptance of these existing treaty provisions, ensuring that the signatories live up to their commitments, and then achieving a consensus on approaches to reduce carbon dioxide emissions are formidable diplomatic tasks facing the international community.

One popular concept being pressed by some developing nations but resisted by most industrial countries is a worldwide tax on emissions of greenhouse gases from energy sources with the proceeds placed at the disposal of poor countries. Such a tax would be sufficiently high to encourage faster movement toward the use of less polluting energy systems. This approach combines the long-standing demands of the developing countries for redistribution of the world's economic wealth, since the developed countries would pay most of the taxes, with international efforts to conserve energy and reduce discharges of carbon dioxide and other pollutants which contribute to global warming.

However, for the concept of an international emissions tax to have any chance of serious consideration by most of the industrialized nations, imaginative ways for using the revenues in the interests of both the developed and developing countries must be devised. The industrialized countries will not simply put more money in the foreign assistance basket. For example, the revenues might be used to offset a portion of the current national contributions to the World Bank and other UN agencies devoted to supporting economic development in the Third World. Conceptually, those countries which are the heaviest polluters of the global commons would have the heaviest burden of supporting economic development of the human resources that populate the globe. In practical terms, the industrialized nations would have an incentive to reduce emissions. Also, they could be encouraged to increase their overall contributions to international foreign assistance through the tax and other means if they are assured that the developing countries, as their commitment, take steps to curtail population growth and introduce cleaner technologies themselves.

Another contentious issue which should be resolved by the time of the Brazil meeting is the international transport and disposal of hazardous wastes. In recent years, the toxic waste transfer industry has become a big international business, as the costs of proper disposal of wastes in industrial nations have increased significantly. By offering developing nations, or more frequently officials in these nations, economic incentives to accept the wastes, Western entrepreneurs have been able to send toxic barges to sea which return to their home ports mysteriously empty. This practice of dumping hazardous wastes in the poor countries may seem abhorrent to American environmentalists, but when countries are in desperate need of money and have unused land, development of a waste disposal industry may make sense to them. Of course a paramount need is to ensure that such an industry is managed responsibly without corruption and does not introduce long-term dangers into the country.

During 1988 a spate of stories appeared in the press about barges laden with wastes that no one would accept. During my visit to Bucharest, Romania, at that time, the story broke that several Romanian ministers were summarily dismissed for accepting toxic wastes from abroad which had resulted in enormous disposal problems along the Black Sea coast. Perhaps these ministers assumed that amid the already heavy pollution in the country, additional wastes wouldn't be noticed. But they were wrong. Highly concentrated toxic wastes pose very different types of hazards than the blankets of pollutants that build up from discharges into the air and water from improperly controlled industrial processes. Politically, coping with domestic pollution is far different than accepting someone else's refuse. Even during Romania's days of dictatorial oppression, toxic wastes broke through the veil of secrecy surrounding governmental corruption. Now, several dozen nations, including the United States, have signed a treaty which forbids transport of hazardous wastes to countries where there is reason to believe environmentally sound disposal may not take place. This treaty requires the written agreement by the government of the receiving country to accept and dispose of toxic wastes in an environmentally safe manner before shipments can take place.

Turning to disposal of wastes at sea, the United States currently belongs to the London Dumping Convention which sets forth limitations on disposal of toxic wastes. This agreement does not prohibit ocean disposal of "normal" municipal waste which is dumped in designated locations along the coastlines of the world. One such site 200 miles off the New Jersey coast which services the garbage scows of New York City has been referred to as "the most disgusting place on Earth" because of the heavy concentration of smelly wastes that are carted there.

International Ecological Security

The international ramifications of environmental issues are growing daily. In 1989 Jessica Tuchman Mathews, writing in the American journal *Foreign Affairs,* tied environmental and related trends to the security of our nation as follows:

> The 1990s will demand a redefinition of what constitutes national security. In the 1970s the concept was expanded to include international economics as it became clear that the US economy was no longer the independent force it had once been, but was powerfully affected by economic policies in dozens of other countries. Global developments now suggest the need for another analogous, broadening definition of national security to include resource, environmental, and demographic issues.
>
> The assumptions . . . that have governed international relations in the postwar era are a poor fit with these new realities. Environmental strains . . . are already beginning to break down the sacred boundaries of national sovereignty . . . . The once sharp dividing line between foreign and domestic policy is blurred, forcing governments to grapple in international forums with issues that were contentious enough in the domestic arena.[12]

These views have been echoed by the president, the secretary of state, and other American foreign policy officials during the past several years. In many American research institutions, ecological concerns are rapidly being accepted as a legitimate dimension of national security by academics and by policy analysts. Foreign policy practitioners throughout the U.S. government struggle to realign traditional concepts of national security within the constraints of established policy directives that have historical roots antedating the current environmental awakening.

For more than 40 years, however, Americans had become accustomed to thinking about national security in terms of military forces—capability to deter a nuclear attack by the Soviets, to repel a land invasion across Central Europe, to quell rebellions in the Third World, and to retaliate against acts of terrorism in the Middle East and elsewhere. National security has traditionally been equated, explicitly and implicitly, with near-term survival by avoiding the ravishes of war—survival of Americans at home and abroad and survival of allies and others who share democratic ideals. A large military–industrial complex is in place, dependent on large defense budgets to ensure their survival through continuous modernization of weaponry and through strengthening logistics capabilities to wage war anywhere—in the sky, on the oceans, and on land.

As highlighted in the *Foreign Affairs* article, this thinking of the past is now giving way to the realities of the future. The crumbling of the Soviet bloc does not mean that the United States is becoming a military ally with the Soviet Union, but it certainly places a different cast on the nature of the nuclear threat and the threat of a Soviet invasion of Western Europe. Our inability to effectively use military forces in Lebanon, Southeast Asia, and Central America has raised great doubts about the viability of national security strategies which pivot around military supremacy even in the Third World.

While the military component of national security obviously remains very important, a challenge of growing magnitude must receive comparable attention—the doubling of the world's population in the next half century and the likelihood of related disruptions of natural systems that sustain human life. During the past 40 years, the percentage of the world's population living in industrialized countries slipped from 40 to 20% and most of the projected population growth will continue in the Third World. As has been the case in the industrialized countries, chemicals will play a central role in shaping the life-styles of these growing populations—in ensuring their food supplies, in providing clothing and consumer goods, and in enabling them to slowly enter the industrial age. How these chemicals are handled will be a pivotal factor in the determination of the future of the planet.

As previously noted, the ability of the United States to influence changes in the world's demography and to mitigate the attendant environmental consequences is highly dependent on domestic policies. The United States is the largest contributor to the most important global ecological problems. At the same time, we are the world's leader in science and technology which can help mitigate the causes of environmental degradation in both the industrialized countries and the Third World. But first we need to have our own house in order if we expect to chart a course for others to follow or even to declare certain courses of development as off-limits.

In summary, the nations of the world are ill prepared to enter the 21st century—the environmental century. The 1990s will undoubtedly be a period of frantic international meetings to improve this state of preparedness. Dozens of international organizations will be vying for leadership roles, and new environmental bodies will be created to raise environmental issues to a higher political level. National leaders from both the rich and poor countries will be pinning their reputations to professed concerns over the global commons. Environmentalists will become statesmen, and statesmen will become environmentalists. Treaties will be drafted, watered down, and eventually signed to limit polluting activities. International foreign aid programs of the World Bank and other UN and regional agencies will be reoriented to promote "sustainable" development—development less reliant on fertilizers and pesticides, development that preserves forested areas, and development that limits the spread of urban areas. However, the critical issue of population growth will remain a controversial topic in Washington and internationally.

Turning to our scientists, they will develop a wide array of new technologies to understand and reduce pollution problems. They will deploy a variety of techniques—from space satellites to underwater research vessels—for measuring the state of the environment. They will continue to search for ways to measure the extent and costs of environmental changes that humans wreak on this planet.

For the United States, a decisive issue will be its approach to national security and to the supporting federal budgets. As noted in an earlier chapter, in the past national security budgets have emphasized support for a triad of nuclear warheads deployed in submarines, bombers, and land-based missiles. America now needs a new national security triad of military strength, economic prosperity, and ecological preservation. National expenditures should be balanced among each leg of this triad. The current distribution of almost $300 billion of federal funds annually for countering military threats, $75 billion (including $60 billion by the private sector) to address concrete pollution problems, and $6 billion for economic cooperation with countries with three-fourths of the world's population is no longer appropriate.

Finally, the struggles between the ideologies of East and West and the economic disparities between North and South will continue during the next decade. But a new struggle between us (the polluters) and us (the victims of our own pollution) will gain international recognition as a crucial security issue of the next century. The nations of the world need to band together as never before if "we" are to be the victors. The international community spends $14 billion each week for support of military forces worldwide. Fourteen billion dollars each year could support a host of new programs in developing countries to help sustain the natural equilibrium not only of the Third World but on a global scale. Which is more important?

9 The Power and Limitations of Science and Technology

If we are going to live so intimately with these chemicals—eating and drinking them and taking them into the very marrow of our bones—we had better know something about their nature and their power.
—Rachel Carson

Science can only give us tools in a box. But of what value are these miraculous tools until we have mastered their cultural and human use?
—Frank Lloyd Wright

Lasers Light Up the Groundwater

In 1984 the housing developments of Henderson, Nevada, just south of Las Vegas were slowly but surely spreading into the vast desert that only a few years earlier had been considered uninhabitable. Green lawns which were sprinkled every day by water from nearby Lake Mead belied the notion that these rock-hard barren areas could not be conquered. Thus, a trip to the "desert" to witness high technology in action took a group of us from the EPA only a few dozen yards from a new row of homes surrounded with cars, pickup trucks, and children's bicycles.

The desert site was less than two miles from the large chemical complex in Henderson which traced its origins to the need for munitions and other industrial supplies to support a wartime economy during the

1940s. Soon thereafter the industrial plants began experiencing the pains of obsolescence. Nevertheless, in 1984 the antiquated facilities continued to provide much of the employment base of the local economy.

Of particular interest, benzene and a variety of other chemicals leaking from large storage containers at the complex had reached the aquifer below us. The residents of the new homes were somewhat concerned about this invasion of the territory underlying their property. However, they had accepted the assurances of the local health authorities that gradual contamination of the aquifer posed no near-term threat.

After rationalizing the situation facing these residents as an inevitable consequence of economic prosperity, my EPA colleagues and I turned to the task at hand. This polluted aquifer which was less than 30 feet below the surface provided a very convenient experimental area for testing the capabilities of new monitoring technology in detecting groundwater contaminants and in measuring levels of contamination.

The visit began with a tour of a large mobile laboratory parked at the edge of the desert. Inside was a dazzling array of laser generators and optical analyzers. A laser light at an appropriate frequency was beamed through a thin tubular fiber encased in a telephone cable. This cable had been inserted into a hole drilled through the desert ground and into the aquifer. The length of such cables can be 30 feet or 3000 feet long since the laser light does not degrade with distance.

The laser beam "sensed" the reactions between chemicals dissolved in the groundwater and a small chemically coated device at the end of the cable called an optrode. The reactions caused a fluorescence or glow of reflected light with the characteristics of the fluorescence dependent on the specific chemical interactions. This reflected light then traveled back up the fiber in the cable. When analyzed in the trailer, the light revealed the presence of polluting chemicals. The light from the returning fluoresced signal was easily separated from the original excitation light sent down the same fiber since it had a shorter wavelength. The key was the optrode which was designed to react with specific chemical pollutants. Different optrodes were used depending on the chemicals likely to be present in the groundwater.

At the time, fiber cables were beginning to take hold as the backbone of the nation's telephone system, and laser-transmission technology had been demonstrated many times over. Fluorosensing of chemicals had also been explored in many facilities, and the determination of chemical signatures from reflected laser signals was an established procedure in military laboratories investigating ways to combat chemical warfare. Still, in the environmental field chemical analysis in well-established laboratories had become the accepted approach for determining pollution levels. The concept of measuring pollutants using reflected light met with considerable skepticism from scientists who were comfortable with the tried-and-true methods of classical chemistry.

The scientists conducting the fiber-optic experiments were from the Lawrence Livermore National Laboratory in California which is best known for its capability to design and test nuclear weapons. They were very excited that they could use technology originally developed for analyzing the results of weapons tests to assess groundwater contamination problems. They beamed with pride as the signals came into the mobile laboratory as expected. They as well as the EPA specialists who were present were very pleased that these measurements were consistent with previous measurements made by the EPA in the same monitoring wells using the traditional method of removing water samples and sending them to a chemical laboratory for analysis.

At the same time, the scientists from California acknowledged the high costs of the experiments which seemed prohibitive in comparison with simply dropping a sampling device down the hole to grab a water sample. While the highly funded nuclear testing programs carried out at the Livermore Laboratory could easily support mobile laboratories that cost hundreds of thousands of dollars, the EPA needed technologies that could be used by tens of thousands of small towns with small budgets and by financially constrained laboratories throughout the country.

Was this just the beginning of a scientific breakthrough in exploring subsurface pollution? The Livermore scientists noted that traditional sampling was plagued with problems. Large holes were needed for the conventional sampling devices. Traditional sampling disturbed the water, and volatile chemical pollutants could escape into the air before the samples reached the surface. Sampling gear became easily contaminated as it was lowered and raised in monitoring wells. The delays in waiting for laboratory results for days or weeks were often very inconvenient.

Fiber-optic technology could avoid all of these problems, according to the scientists. The telephone cable could be inserted into a very small hole and left in place for many months. Measurements could be made instantaneously. Costs would surely fall as the technology became widespread.

Within 12 months, the potential of this technology began to emerge. The scientists from the Livermore Laboratory met me in San Francisco in early 1985 where they briefed the EPA Regional Administrator on the progress of the experiments. They brought along 35-millimeter slides which showed their field operations, but they left the mobile laboratory at home. Instead, they arrived in San Francisco with their new "laboratory." All of their equipment had been miniaturized, and a small suitcase now replaced the mobile van. A light bulb and filters had been substituted for the laser generator, and highly compact electronic devices were used for the analyses. The equipment was battery powered. The cost of the suitcase was $2500, but it was one of a kind. Mass production certainly would reduce the cost further.

In the years that followed, this environmental application of fiber-optic technology has continued to develop, and its application remains of great interest to many environmental agencies. However, as with any new technology, considerable time is needed to refine both the equipment and the procedures for using the new devices. The accuracy and reliability of the equipment must be demonstrated many times in the field so that the data it obtains will stand up to close scrutiny, even under rigorous cross-examination in courtroom settings. In some applications such as repetitive screening for leakages around underground storage tanks for a few well-known chemicals, the value of this technology seems obvious. However, in other situations such as providing definitive evidence on the presence of all possible chemicals in groundwater near a chemical waste site, detailed laboratory analyses of water samples will retain their importance.

Fiber-optic technology is just one example of the many new technologies that can reduce the costs and improve the timeliness of assessments of subsurface environmental contamination. Now let us turn to an example of an older technology with great potential for assessing environmental problems on the surface of the Earth and in the atmosphere.

Finding Pollutants from Airplanes and Satellites

For more than a dozen years aerial photographs have been very important in the national effort to search out abandoned waste sites and poorly managed sites and then to assess the environmental damage near those sites. Photographs provide concrete evidence of irresponsible behavior by chemical dumpers and by operators of waste disposal sites who try to avoid disposal regulations. Such photography is usually obtained from low-flying aircraft although high-altitude photography and even satellite photography have also been very helpful in assessing the destructive impacts of man's chemical litter.

During the 1980s other types of remote sensing technology from aircraft played an important role in our environmental assessment programs. For air pollution, laser systems called "lidars" have been extensively deployed in aircraft, especially over southern California, to study pollution plumes of particulate matter. They complement ground-based air-monitoring instruments which are not able to provide rapid measurements over very large areas. Airborne measurements are critical to understanding the long-range transport of air pollutants. They can reveal the structure of urban haze as well as more definable plumes. They are helpful in linking traces of particulates and other air pollutants with specific pollutant sources. Also, the measurements can be used in mathematical models of problems plaguing large metropolitan areas.

Extensive research efforts have also concentrated on providing airborne laser systems which not only track particulate air plumes but in addition can detect the presence of invisible polluting gases, particularly ozone and sulfur dioxide. Measuring the presence of such gases will help improve our understanding of the extent and causes of pollution problems which are often independent of the problems caused by aerosols of particulates. However, the technical difficulties are formidable. Even though the volumes of pollutant gases which are discharged into the atmosphere can be very large, the gases disperse rapidly. At any given time the concentrations at a specific location are small and difficult to detect. Also, many different gases are of interest to scientists, but a laser sensor is usually tuned to detect only one or two different types of chemical molecules at a time. Perhaps in the future they will be able to detect larger numbers of pollutants simultaneously. Nevertheless, this technology already offers important capabilities.

As we turn to water pollution, we see that laser systems have again proved their worth. Airborne sensors can detect elevated or depressed levels of chlorophyll *a* which is associated with growth of algae in lakes and rivers. Many algal profiles of water boundaries have been determined using these systems. High levels of algae cause foul-smelling water and generally unpleasant aquatic conditions while abnormally low populations of algae may indicate the invasion of substances which are toxic to algal growth and possibly to humans as well. Also, as previously noted, algae can be an indicator of the presence of nutrients needed to sustain fishery resources.

Airborne laser systems also measure the presence of dissolved organic carbon in surface waters which is frequently linked with either man-made or natural polluting substances. Although such substances are seldom highly toxic in low concentrations, they often act as carriers for toxic pollutants. Dissolved organic carbon in raw sources of drinking water raises special concerns since during the chlorination process traces of organic compounds might be converted into carcinogenic chemicals.

Overshadowing laser systems for investigating pollution on the ground or in aquatic systems, however, has been the rapid spread of multispectral sensing technology. This technique discerns abnormalities on the surface of the Earth by measuring reflections of the sun's light in different wave bands. Two decades ago, the U.S. space program dramatically demonstrated this technology. Since that time, both satellite and aircraft multispectral sensing have become popular components of environmental programs. For example, the EPA has used systems mounted in aircraft to determine the biological conditions of Flathead River, Montana; to assess the influences of warm and cold springs in Mono Lake, California; to analyze septic system failures in Windemere, Minnesota; to locate underground coal mine fires in Monarch, Wyoming; and to classify vegetation in Big Bend National Park, Texas. The EPA and other organizations have used satellite systems to trace ocean currents, to delineate ecological destruction from shoreline erosion, to identify areas of forest blight, and to reveal downstream impacts of runoff of toxic metals from mining areas.

However, in 1984 scientists at the EPA's Las Vegas laboratory found resistance in trying to introduce this modern technology into an environmental program which had already been conceived along traditional lines. The EPA headquarters in Washington had called upon the laboratory to direct the sampling of a large number of lakes, using helicopters and mobile laboratories, in areas of the Midwest, New England, and the Southeast which were believed to be particularly vulnerable to acid rain. The Agency's primary interest was finding out the extent of the acidity of the lakes. Measurements of other chemical properties of the lakes in addition to acidity would contribute to understanding *why* the lakes had or had not become acidic.

About 200,000 lakes are located in these vulnerable regions. The EPA selected 2000 as a reasonable number for sampling within the cost constraint of about $15 million for the program. For sampling purposes, the statisticians classified the lakes according to their geographical locations, their altitudes, and their sizes. However, laboratory scientists were concerned whether such a small sample would truly be representative of all the lakes since in many ways each lake is unique. Statistical extrapolations from 2000 to 200,000 lakes would be uncertain at best.

Therefore, the laboratory proposed as a complement to the planned sampling of 2000 lakes the use of multispectral sensing data that could be obtained from satellites which were already in orbit. This information would permit examination of the conditions of many more lakes where the helicopters would not land. For those lakes where the helicopters would land, correlations of monitoring data from the classical chemical measurements and from multispectral sensing would enable scientists to "calibrate" the sensor data. They could then infer from the satellite imagery some of the conditions of the lakes which were beyond the range of the helicopters.

Unfortunately, many skeptics within the EPA were not prepared to support this approach even on a limited experimental basis. They argued that there were no ways to measure the acidity of the lakes directly from sensor data since the proven sensor "signatures" of lake conditions (e.g., turbidity, clarity, algae) are not directly related to acidity. However, tests under other programs had shown a high statistical correlation (80%) between the actual acidity and inferences of acidity from analyzing combinations of other less certain signatures that could be measured directly from satellites. Also, the laboratory argued, the multispectral analyses were not intended to displace the conventional statistical extrapolations from the direct water sampling. Rather, they would help determine the uncertainties in the extrapolations from the acidity conditions in the sampled to the conditions in the unsampled lakes.

Finally, the most persuasive argument, so we thought, was that the sensing and analyses were relatively cheap (a few hundred thousand dollars) since the satellites were already in orbit. Also, the program would provide the EPA with invaluable experience for future investigations of lake conditions as this technology continued to advance. Nevertheless, the EPA simply was not ready to enter the modern age of remote sensing, and the Agency did not agree to provide funds for the use of multispectral scanning. The assessments were to rely only on the direct sampling of the 2000 lakes. Since that time, fortunately, many organizations including the EPA have finally embraced this type of satellite sensing as an important tool in assessing the conditions of lakes.

Indeed, during the past several years global remote sensing systems have been heralded throughout the world as the backbone of the international effort to assess the environmental state of the planet, including impacts of acid rain. Every international plan for protecting the global ecology calls for expanded systems of satellites and aircraft to assess regularly the state of the stratosphere, the atmosphere, and land and water surfaces. The presence of greenhouse gases, the condition of the ozone layer, and the denudation of forest resources are of special concern. Scientific studies of air–sea interactions as measured by satellite systems are also improving our understanding of the capacity of the oceans to absorb carbon dioxide.

The list of potential environmental applications of remote sensing systems is very long. Some applications such as photographs of waste sites relate directly to assessing the impact of chemicals on ecological resources. Other applications such as measuring changes in vegetation and land use can provide indirect indications of the effects of chemicals on the ecology. In numerous ways, remote sensing not only saves time and money, as compared to other data collection techniques, but it also provides data that cannot be collected in any other manner.

The Achilles' heel of remote sensing has always been the limited capabilities for fully using the collected data. More powerful and more friendly computers are a critical part of the answer. Unfortunately, only a small portion of the potential users of ecological information are computer literate, and in many developing nations which are vitally interested in environmental conditions of the planet, computer skills are at a particularly low level. A program to promote remote sensing skills needs to be undertaken without delay at home and abroad. Also, there is no better topic than global ecological change as a focal point for advancing general computer literacy.

Technology and the Greenhouse Gases

While scientists seek improved capabilities to assess environmental problems, engineers are looking for approaches that will reduce the amount of man-made chemicals reaching the environment. As previously discussed, the by-products of fossil fuel combustion, which are major contributors to the greenhouse effect, pose a particular type of hazard.

A large number of technological approaches to reducing the adverse impacts of energy systems on the environment, and particularly curtailing the emissions of greenhouse gases, are in various stages of development and use. Some of these technologies can provide new opportunities for conservation practices by the consumers of energy and thereby reduce energy demand. Others can stimulate the introduction of substitute manufacturing processes or products with reduced requirements for energy. Developing improved efficiencies in the generation and distribution of electricity is also an important technical objective for conserving energy. Fuel switching to increase reliance on less-polluting energy sources depends on technological successes in developing cheaper and reliable alternatives. Finally, new techniques can be used to improve pollution control systems within energy-producing plants. A few of these technological opportunities are discussed below.

Conservation Practices by the Users of Energy. Many techniques are available for designing thermally efficient residential and commercial buildings. They begin with the basic architecture of the buildings. Also, new types of construction and insulating materials, improved window designs and air circulation systems, and even "smart houses" with electronically controlled heating and lighting systems are becoming more important. Very minor engineering modifications of existing heating systems can often improve energy efficiency both in building complexes with central heating and in structures which have their own heating systems. Finally, the energy savings associated with the advent of the heat pump as a heating and cooling device in many homes and commercial buildings have stimulated keen interest in recent years.

Similarly within industrial settings, improvements in energy efficiency depend on a variety of approaches, and frequently relatively minor engineering adjustments of the current manufacturing processes can have significant payoff. Adjusting operating procedures to take advantage of lower-cost electricity during off-peak hours sometimes makes good business sense while saving energy for the geographical area. Some industrial processes, such as metallurgy, are highly energy intensive. Even small-percentage energy savings through more careful control of the temperatures and timing of the manufacturing processes can meaningfully contribute to less pollution.

Improved Efficiency in Electricity Generation and Distribution. Fluidized-bed combustion and integrated gasification–combined-cycle technologies have been shown to provide great advantages in capturing heat that would be otherwise lost from coal-fired power plants. With fluidized-bed combustion, large fans keep powdered coal suspended in midair so it burns cleaner with less energy lost. In integrated gasification plants, coal is turned into a gas, removing sulfur in the process. The gas is then used to run two turbines, one powered by the hot combustion gases and the other by steam.

In recent years natural gas technology has developed rapidly. Combined-cycle systems are analogous to coal gasification systems, relying on both a natural gas-fueled combustion turbine and a heat recovery steam generator and steam turbine. Also, gas can be used in combination with coal: gas–coal co-firing involves the introduction of natural gas into the primary furnace combustion zone of a pulverized coal boiler. This process has the benefit of reducing emissions of nitrous oxides and sulfur dioxide while improving overall system efficiency.

Let us turn to hydropower. While environmental objections frequently thwart the building of new dams, adding and refurbishing turbines at existing sites can often increase output and efficiency at relatively low costs.

Related to improved electrical efficiency is the rapid spread of "co-generation" systems. These systems are built around power plants which generate electricity. While electricity is being generated, either the very hot water or the steam which is produced during the electrical generation cycle is sent through pipes into local residential areas or even into large metropolitan districts as the basis of their heating systems. In some experimental systems, heated natural gas is used both in the power plants and in the heating systems.

Fuel Switching. During combustion, natural gas emits 30% less carbon dioxide than does petroleum and 50% less than coal for equivalent energy production. Therefore, natural gas offers a very attractive near-term fuel alternative in those parts of the world where gas is no more expensive than oil or coal. In some areas where distribution systems are in place, greater reliance on natural gas can take place quickly. In others, the capital costs of distribution systems will undoubtedly inhibit the introduction of new natural gas power plants. In the long run, limitations on the amounts of natural gas which are available will force the use of much of the gas for activities of higher value than electrical generation. For example, natural gas is an important raw material for production of industrial chemicals. Thus, the use of gas for generating electricity will be increasingly confined to localized areas with large supplies.

With regard to renewable sources other than hydropower—such as solar, wind, biomass, and geothermal—the uncertain economics, the need to demonstrate the capability of plants to operate reliably for 25 to 30 years, and limitations on appropriate geographic locations for these technologies will likely constrain their use in the next 50 years. Of course, biomass plants such as those relying on manure in California and on wood chips, tree trimmings, and construction materials on Staten Island should be encouraged even though they will not be large energy contributors. Further research on the use of methanol and other alternate fuels for motor vehicles is important even though the energy content of one gallon of these substitutes is much less than the energy content of one gallon of gasoline. The blending of alternate fuels and gasoline is a particularly attractive approach. Also of special interest is the steady and significant progress in photovaltaic technologies, both in efficiency and in cost. All of these technologies can become minor but nevertheless important contributors to the energy mix during the next century.

Finally, as we have seen, nuclear fission has the potential of being a major energy contributor in the decades ahead. Current designs of inherently safe reactors which shut down in emergency situations through the laws of physics rather than through reliance on human intervention are gaining increasing support from governments and the public. Meanwhile, improvement of maintenance and operating procedures for the current generation of nuclear reactors is urgently needed around the world to prevent accidents that could further jeopardize the future of nuclear power.

Improved Environmental Control of Energy By-products. Several types of pollution-reduction devices have been installed in power plants for many years, and others are under development. Catalytic devices route waste gases through chemical compounds which convert some pollutants to less harmful substances. Many plants have demonstrated that emissions of nitrous oxides, for example, can be reduced by 80–90%. Wet lime or spray dry flue gases can be effective for desulfurization of wastes from coal. Still, additional research is needed to develop even more effective controls for emissions of carbon dioxide.

With regard to mobile sources, most of us are familiar with removal of nitrous oxides emitted by motor vehicles using catalytic technology. In California, as noted, attention is now directed to warning devices which alert the driver when the catalytic controls are not working properly. Meanwhile, the tightening of emission standards is forcing further refinement of catalytic converters not only in the United States but also in many other countries.

Less Energy-Intensive Products. Fuel efficiency of motor vehicles has been a long-standing concern in the United States and other countries. Steady progress is being made to reduce energy requirements of the vehicle fleet. Lighter-weight materials which do not sacrifice safety and improved combustion systems continue to be key aspects of this effort.

A second example of large potential energy savings is the current effort in many countries to reduce their reliance on agricultural pesticides and fertilizers that require high usage of energy during their production. Not only is the consumption of fuels high, but production processes require relatively clean fuels (e.g., petroleum and gas). These fuels are then not available for other uses. Biotechnology and improved tilling practices, which are discussed below, are among the rapidly advancing techniques which hold promise for reducing the need for agricultural chemicals and the attendant drain on relatively clean energy sources.

Farmers Seek Environmental Acceptance

Continuing with the agricultural theme, the Texas Commissioner of Agriculture recently gave the following gloomy assessment of the state of American agriculture:

> In recent years, we've begun to run up an agricultural deficit. Skyrocketing production costs have outstripped farm income. Once thriving farm families sometimes cannot put enough food on the table for their own needs. Increasing concentration in food production and processing operations is reducing the options available to consumers, raising prices, and threatening food safety. Our topsoil is being lost at rates comparable to those of the Dust Bowl of the 1930s: parts of Iowa are losing two bushels of soil for every bushel of corn produced. Our water supplies are becoming contaminated with pesticides, and public health officials are issuing new warnings on the risks of exposure to pesticides for agricultural workers. All the while, pests are growing more resistant to our poisons.
>
> The great decline in American family agriculture is the direct result of a system that benefits the corporate farm at the expense of the family farm; that too often stresses cultivation of a single crop to the detriment of diversified agriculture; that is geared for high volume, low price exporters and megacorporate processors rather than for actual demand and local processing; that breeds dependence on expensive synthetic chemicals rather than the replenishment of natural resources.
>
> It is the result of a system that locks farmers into debilitating pesticide and fertilizer cycles that deplete the real value of American farms. The land itself becomes addicted to chemical fertilizer, becoming less productive and losing its value. Farmers find themselves spending more in order to put more chemicals into the land each year while getting ever-diminishing returns.[1]

Echoing these concerns, a 1989 report of the National Academy of Sciences concluded that "alternative agriculture systems are economically feasible but are discouraged under federal commodity programs which have a very short-term focus. Alternative systems can sustain production more effectively over the long term. They can minimize harmful environmental side effects that adversely affect people and ecosystems off the farm. Alternative agriculture does not reject conventional practices but more deliberately takes advantage of naturally occurring beneficial relationships—relationships between pests and predators and between nitrogen cycles and plant growth. "Successful alternative farmers do what all good managers do—they apply management skills and information to reduce costs, improve efficiency, and maintain production levels," stresses the Academy report.[2]

According to the National Academy of Sciences, federal agricultural policies work at cross-purposes to the nation's environmental policies and discourage environmentally compatible farming systems which could, for example, rely more heavily on crop rotations and soil conservation while reducing applications of pesticides and fertilizers. The report concluded that even in the absence of federal support, alternative agriculture can be productive and profitable. There are many well-managed alternative farms which use fewer agricultural chemicals—practices that do not decrease but in many cases increase per-acre crop yield and productivity of livestock. The evidence assembled by the academy is persuasive that wider adoption of alternative systems would result both in economic benefits for farmers and in environmental gains for the nation.

In addition to urging modifications in federal subsidy policies which encourage planting of single "program" crops to attain the highest possible immediate yields, the report of the Academy called for expanded research in this area. Improved scientific understanding can broaden the types of proven approaches to agriculture. The academy notes that among the best known alternative farming methods that deserve much greater emphasis are:

- Crop rotations that mitigate weed, disease, insect, and other pest problems; that increase available soil nitrogen and reduce the need for purchased fertilizers; and that help reduce soil erosion.
- Integrated pest management which can cut the use of pesticides by greater reliance on other techniques such as using biological agents to control pests, introducing pest-resistant crop vari-

eties, and "scouting" to assess when pest problems are so severe that pesticides are the only alternative.

- Animal disease prevention through health maintenance rather than through the routine use of preventive antibiotics which can over time influence the quality of animal agriculture and can find their way into the environment.
- Genetic improvement of crops to resist insects and diseases and to use nutrients more effectively.[2]

This current questioning of conventional agricultural practices brings into focus several personal experiences which should have signaled flaws in the nation's approach to agriculture.

When as a young boy I first visited my uncle's farm in Illinois, I was thrilled to fly in his private plane. We had an aerial view of hundreds of square miles of cornfields where he and his neighbors applied modern science and technology to achieve record yields. His management approach included slicing the runway for the plane through the middle of his richest cornfield and also not wasting time in trying to reclaim marginal lands at the edge of his property. The runway as well as the marginal lands became part of the soil bank program. He received a lucrative subsidy for not planting corn in what appeared to be the most fertile part of his property which had been reserved for the runway while he used large amounts of chemicals to increase production to the limit on other parts. This type of subsidy which promoted the use of private airplanes and destroyed any incentive for rehabilitating marginal lands simply didn't make sense to me.

Fifteen years later as a student studying the use of water from the Colorado River, I was amazed to see how the southern California desert, with topsoil of only a few inches, had been brought to life through the wonders of irrigation. But state agricultural officials dismissed the buildup of salt along the banks of the irrigation ditches, the polluted state of the water discharged off the farms, and the declining levels of the rivers and lakes that fed the irrigation system. The folly of this emphasis on production at any price, even in the middle of the desert, is now being exposed as more and more Americans move into these desert regions, depleting limited water supplies and fouling the desert environment.

Twenty years later while visiting the University of Nebraska in

Lincoln, I was very impressed to learn how university professors drawing on American science and engineering knowledge had dramatically increased crop production not only in the United States but also in Latin America and other distant regions. However, they themselves were questioning the soundness of their approaches. They were particularly concerned that the underground aquifers throughout the state of Nebraska, including aquifers under their own property, were rapidly deteriorating as pesticides and fertilizers drained from the fields into the ground water. From all indications in some areas the pollution levels had exceeded the levels of reversibility.

For many decades American agriculture has had no rival. Scientists, extension agents, and farmers throughout the nation have shown the world how modern science can bring bountiful harvests to feed an entire population with enough left over to feed other countries as well. Indeed, supermarkets are a cornerstone of American affluence. Now our scientists, our agricultural community, and the nation are challenged to maintain the harvest while changing those farming trends which are threatening the environment.

The Promise of Biotechnology

Biotechnology based on recombinant DNA is an important scientific breakthrough that should help mitigate some of the adverse side effects of farming practices on the environment. Experiments have demonstrated the feasibility of genetic engineering of crops to increase their internal resistance to pests, to weather, or to herbicidal chemicals designed to kill weeds that threaten the crops. At the same time, the newly induced traits of these plants enable them to produce better and higher crop yields. For example, genetically modified bacteria allow strawberries to withstand frost; molecular manipulations protect cereal crops from insects; and a specially modified gene produces firmer tomatoes.[3]

The traditional methods of selective breeding of hybrid crops which can withstand adverse environmental conditions while producing high yields require many crop generations before the genes with desirable characteristics are established as part of the permanent genetic makeup of the plants. Even after years of trials there can be no guarantees that the crossbreeding will be successful. Biotechnology truncates the development cycle. Another important benefit is that only the specific genetic traits that are desired will be incorporated into the plants. Genetic engineering has a degree of precision that cannot be achieved with breeding. In addition, DNA technology allows foreign genes to be introduced into plants whereas crossbreeding relies only on genes that are already present in the species used in the experiments.

Biotechnology has applications with important environmental benefits in fields other than agriculture. For worldwide energy, converting biomass to raw chemicals through biotechnology may in some instances be a feasible alternative to petroleum as a raw material for industry. Petroleum is of course a finite resource, and the extraction and refining of crude oil is surrounded with potential environmental problems. In contrast, renewable sources of biomass stand underutilized—starch and cellulose from corn and grain, forest products, and even some municipal wastes. In concept, biotechnology processes using DNA technology can enhance the enzymatic actions of microorganisms which break down biomass into forms that can be converted into industrial chemicals. The technology is not yet in hand, but in the future this approach may contribute toward reducing our dependency on oil.

With regard to pollution control, biotechnology may be able to mimic certain natural processes that feed on toxic chemicals. For decades wastewater treatment plants have used microorganisms to help eliminate solid particles. Genetically altered strains of microorganisms offer the promise of enhancing these processes, and perhaps they will eventually be able to degrade hazardous wastes into harmless chemicals. As one example, laboratory experiments have shown that a common rot fungus is able to degrade PCBs and DDT. Investigations are under way to determine if such a fungus, bioengineered to enhance its degradation capabilities, could be spread over a hazardous waste landfill to destroy toxic chemicals.

Microbes are commonly used to leach metals from mining ores and wastes, particularly in the copper industry. Again DNA technology may be able to produce more efficient microorganisms which would flourish in the acidic conditions of mines. In the petroleum field, experiments are being conducted to determine whether microorganisms or microbially produced substances could enhance the recovery of oil that cannot be easily extracted by conventional methods. As an example, bioengineering has produced a fatty substance that reduces the viscosity of crude oil, thus allowing thick oil to be pumped from the ground.

Beginning in the mid-1970s many scientists became concerned that a "superbug" produced by biotechnology might escape into the environment, reproduce in an uncontrolled manner, and create havoc. To prevent such a possibility, federal agencies have promulgated regulatory guidelines. Also, many individual research institutions have established scientific advisory committees to monitor the use of genetically engineered organisms in the environment.

Scientists continue to caution against the release of genetically altered microorganisms. They cite examples of disastrous effects that resulted from the introduction of new species into the environment. Dutch elm disease, gypsy moths, and starlings were all introduced into new ecosystems. With no natural predators, these organisms upset the ecological balance and reproduced in epidemic proportions. Could genetically engineered microorganisms, like nonnative species, have adverse effects of unknown proportions as they reproduce and multiply without the possibility of recall? Or is ecology simply being distorted and portrayed as a subversive science which threatens to thwart the advances of modern technologies?

The National Academy of Sciences pointed out in a 1989 report that the techniques used to genetically engineer organisms are not intrinsically dangerous. The knowledge gained from centuries of plant breeding and decades of modifications of microorganisms, combined with a good understanding of potential release sites, is often sufficient to permit evaluation of the danger of introducing a genetically modified organism into the environment.[4]

The academy notes that the potential for "weediness" has been cited as a major environmental risk posed by introduction of genetically modified plants. Weediness means the ability of a crop to go out of control and to become a weed species itself or, alternatively, to hybridize with wild plants to produce weedier progeny. However, the academy pointed out that the likelihood of enhanced weediness is low for genetically modified, highly domesticated crop plants. The academy also observed that microorganisms are more prone to spontaneous mutations than are plants or animals, but these risks can be evaluated scientifically. One way to confine introduced microorganisms to target environments is to introduce "suicide genes" into the organisms—such as temperature-sensitive genes which ensure that the organisms cannot survive outside the target environment.

Products of biotechnology are gradually becoming ready for the marketplace. The U.S. Department of Agriculture and the EPA have granted about 100 permits for field testing of genetically modified organisms. During the next decade we can expect to see a variety of applications of this technology which holds considerable promise for displacing other approaches that engender many environmental problems.

Technological Opportunities and Social Engineering

The foregoing examples of scientific and technological opportunities to assess and abate the spread of chemicals in the environment begin to explain why many scientists are optimistic that worrisome trends in air and water pollution can be reversed. Hundreds of other recent innovations are also of great interest in the searches for "environmentally soft" industrial processes and products and for technologies which can help identify existing environmental problems and predict future ones. Many current research and development efforts are pointed in these two directions.

Industrial, agricultural, and other commercial interests are intensifying their efforts to find production technologies that will generate fewer hazardous by-products and still remain economical. Our government can encourage these efforts by setting standards that must be met, such as requiring the use of the best available technology that minimizes toxic wastes in manufacturing facilities and calling for the best management practices in the countryside that reduce agricultural runoff. The government can levy taxes on gasoline consumption and on chemical waste streams, and it can provide financial incentives in the form of subsidies or tax deductions for environmentally sound practices in the home and in business. The government can also ensure customer awareness of the characteristics of products with environmental implications, such as gasoline consumption traits of new cars which are conspicuously posted by every car dealer and the hazards of chemical shipments which are indicated on warning labels.

Still, the primary burden for softening the nation's technological base must rest with the private sector. Our technological base is enormous and has become fully integrated with the nation's workforce and the nation's financial institutions. Change will not come easily, but it must come.

With regard to research to improve the country's capabilities for assessing environmental problems, the federal government has traditionally taken the lead. The research budgets of government agencies reflect this responsibility for investigating and developing new and improved assessment technologies and techniques. While industry of course has always been expected to be alert to the problems it may be causing, too often liability suits have been needed to capture the attention of industry even to apply well-known assessment and mitigation approaches. Similarly, in the agricultural sector, commercial interests frequently have preferred not to investigate the obvious side effects of increasingly intensive farming methods.

Historically, a few large companies have long supported significant research programs in the fields of environmental assessment and control. In recent years, the number of companies financing such activities has increased dramatically. As many companies are forced to pay the bills for cleaning up environmental problems, they usually prefer to rely in the first place on their own scientific staffs for advice rather than simply being directed by the government as to how they should move forward with remediation efforts. These staffs in turn become internal company lobbyists for more aggressive environmental programs. Also, many engineering service companies see ever more lucrative governmental contracts for environmental cleanups on the horizon, particularly within the Superfund program; and they are rapidly expanding their technical wherewithal to compete for these contracts.

A flash point for heated debates in Washington over the adequacy of research and development activities to address environmental problems is the size of the EPA's research budget. In 1980 the budget was about $350 million. Ten years later it remained at that level despite the growing array of environmental problems, the dramatic increase in the overall operating budget of the Agency, the ever-expanding commitments of the Agency to provide technical support for the states, and inflation. Many of the scientific uncertainties that confronted the EPA in 1980 loom larger than ever in 1990. We know little about the neurotoxic effects of chemicals. The resiliency of ecological systems to rebound to their original states after being temporarily stressed by chemicals is now little more than a theory. Extrapolating effects of chemicals on laboratory animals to effects on humans continues to baffle scientists from many disciplines. The feasibility of detoxifying concentrated wastes and residues by spraying them with selected chemicals has yet to be demonstrated outside the laboratory.

One hopeful sign in Washington is the strong budgetary support that is now being directed to understanding global warming. The federal research program calls for seven agencies to devote $1 billion in 1991 to supporting research in seven categories. They are climate and hydrological systems, biogeochemical dynamics, ecological systems and dynamics, the history of the Earth's climates and environments, human interactions with global systems, solid earth processes, and solar influences. When these efforts are combined with similar research undertakings in other countries, the government is well on the way in its scientific commitment to clarify the uncertainties of one of the most ominous problems confronting all nations.

In short, both the environmental regulatory agencies and the research agencies of government influence technological developments throughout the nation. However, these agencies are not well equipped for the task. Neither they nor anyone else comprehends the profound changes that will reverberate throughout society as conventional concepts of the economic viability of commercial endeavors are modified in response to environmental concerns.

For the past two decades our nation has been engaged not only in environmental engineering but also in social engineering. We are attempting to change the very basis of our life-styles. The changes demanded of our population thus far have been modest in comparison to what lies ahead if America is to continue to enjoy a tolerable environment.

During these 20 years, the president has had a Council on Environmental Quality which has been bogged down in the minutiae of regulatory details which are the bread and butter of the regulatory agencies. In 1990 it took a step toward executive leadership in its publication of recent environmental trends. However, a far broader vision is needed, a vision that goes beyond pollution to life-styles. What better task could there be for an organization at the highest level of government than to provide a perspective of the future of a nation in responding to environmental stresses of unprecedented magnitude? Such a perspective would provide invaluable guidance in how best to target resources for research, including research on social as well as physical processes, as the nation girds for the long haul.

10 Living with Toxic Risks

Nothing so fair, so pure, and at the same time so large as a lake, perchance, lies on the surface of the earth.
—Henry David Thoreau

To waste, to destroy, our natural resources, to skin and exhaust the land instead of using it so as to increase its usefulness, will result in undermining in the days of our children the very prosperity which we ought by right to hand down to them amplified and developed.
—Theodore Roosevelt

The United States does not now face an environmental crisis . . . . Looking ahead, however, there is a set of complex, diffuse, long-term environmental problems portending immense consequences for the economic well being and security of nations throughout the world, including our own.
—Administrator of the U.S. Environmental Protection Agency William Reilly

When EPA officials become discouraged over progress in cleaning up America, they often console themselves by saying, "We're doing better. We're only feeling worse." Many of us surely are feeling worse as the experts uncover new sets of environmental problems each year. More sensitive analytical chemistry techniques reveal previously unrecognized traces of man-made chemical contaminants in food and water and in the air of our nation's living rooms. As bulldozers excavate below the ground's surface to prepare building foundations or simply to level the land, they unearth chemicals which were discarded or spilled and hidden with dirt. Some of these chemicals have slowly spread like unbounded oil spots. Around the globe, evidence of human wastes is appearing everywhere, from toxic metals in the purportedly pristine Bering Sea to acid deposition in uninhabited deserts. Finally, we need no reminder about the hot summer of 1988 which some believe surely portended a future of hothouse living under a blanket of air pollutants.

But are we doing better? Washington has enacted extensive environmental legislation, and all state capitals are issuing regulations that seem to touch every facet of our lives. In many regions of the country, rivers have come back to life and air pollution levels have declined. The government has restricted the use of a number of potent pesticides and other dangerous chemicals. Moreover, a number of U.S. attorneys, supported by many investigators often called eco-cops, work full time tracking down violators of environmental regulations. Every major manufacturing company now claims an environmental conscience. Ecological issues have risen to the top of the agendas of summit meetings of heads of state. Green organizations are on the alert in almost every country.

Yet the gains in mitigating pollution pale in comparison with the potential severity of environmental problems. Even with the anticipated reductions in emissions of air pollutants, in some areas the remaining levels will still cause harm to people and to ecological resources while steadily adding to the overall contaminant burden placed on the Earth. Water pollutants may be temporarily out of sight as they cling to sediments at the bottom of rivers only to resurface when the sediments are disturbed. Pollutants that have been accepted as safe at low levels may not be totally benign. While they may not cause recognizable diseases or disorders, some have very subtle effects on genetic systems which in time can change human or ecological characteristics. Indeed, we can no longer limit environmental concerns only to the "adverse" effects of exposures to chemicals entering the environment. "Adverse" defies definition, and all types of chemical side effects must now receive attention.

In terms of regulations and cleanup activities, we as a nation are doing much better than in 1970, in 1975, in 1980, or in 1985. But we're not doing well enough for the 1990s. In homes, consumption patterns remain wasteful and in many ways incompatible with environmental improvement. If Americans really need to use two billion disposable razors each year, communities must learn to handle plastic wastes. Americans now have so many automobiles that the entire population of the United States can ride in the front seats. If Americans insist on driving 10 to 20 miles to work, gas guzzlers must become a historical artifact. If our megacities continue to grow, support for fume-free mass transit must find a place in city budgets.

In undeveloped countries, population growth has skyrocketed with attendant pressures on the land, forests, and rivers. This population explosion can no longer continue unchecked. If the Chinese begin to use their low-grade coal on a massive scale, the skeptics who doubt the likelihood of future greenhouse summers will rapidly lose their skepticism.

Fortunately, national governments around the world have finally awakened to the severity of environmental problems. They recognize that human life is fragile and that survival is at stake. We have no alternative but to do better in the 1990s.

* * *

How much should the United States invest now to protect its ecosystems and its people from eventual destruction? How does society decide how much is warranted?

Unless human activity stops altogether, the environment cannot be free of man-made contaminants. In all areas of environmental protection, the core issue is how clean is clean enough. When should an aquifer be abandoned as a source of drinking water? When should fishing grounds be closed? How deeply should leaking waste sites be excavated? When do traces of toxic chemicals in the air reach an unacceptable level?

Human factors compound the difficulty confronting governmental bodies and individuals in reaching day-to-day environmental decisions that determine the future of the nation and the planet. Decisions must be made despite the uncertainties of science and the uncertainties as to how society will react to new ground rules for living. Some government officials are uncomfortable with uncertainties and too often prefer to procrastinate when faced with doubt about environmental hazards and economic costs. Many scientists are reluctant to acknowledge uncertainties in their judgments lest their views be dismissed. At the same time, the public is no longer hesitant to question the assertions of experts, and lawyers seize on uncertainties to discredit their opponents.

The preceding chapters have presented a few signposts that I hope will be helpful in charting the many routes that America should follow. Our society can successfully navigate the shoals of environmental hazards only with a rudder of boldness since time is too short to accommodate all the social biases of the past. Now, national debates over our environmental future must give greater attention to core issues that were easier to avoid in previous years: the use of federal lands as sites for chemical disposal, the intrusion of federal and state authorities into local zoning decisions, the acceleration of the development and use of nuclear power, the introduction of higher gasoline and energy taxes, and the commendation as well as the condemnation of industry for their environmental activities. The EPA must have the backbone to resist political pressures to try to solve every problem at once and thereby solve none. For example, it makes much more sense to thoroughly clean up a limited number of chemical waste sites each year than to find partial solutions for every waste site in order to satisfy the demands of local political constituencies—partial solutions that require additional solutions in future years.

America will founder on the rocks of self-destruction if the hands on the tiller of environmental policies steer by concepts that insist on the primacy of near-term economic betterment over environmental quality and concepts that are wedded to how it was rather than how it must be. Every American has a stake in the environment, and every American will have some influence on the quality of life on the planet. We will all decide our environmental future. Elected leaders, appointed officials, economic barons, and articulate journalists may have the most visible votes. But 250 million Americans are shareholders in this enterprise, and every day they are redefining the meaning of proxy votes.

* * *

How can the entire population effectively participate in determining the future environmental quality of the nation when even the experts disagree on almost every major prevention or abatement program that is proposed?

Our system of "modern" education will be tested as never before.

Parents must stress environmental literacy in the home at the earliest age. Learning about ecological resiliency and sustainable development must continue in the schools, the universities, and beyond. Teachers and professors, camp counselors and park rangers, museum and exhibit directors, television and press commentators, and friends and colleagues all play important roles in this educational process. But self-instruction will be critical as Americans increasingly experience the impacts of environmental degradation and are confronted with environmental rules and regulations which constrain personal behavior.

Within the formal system of education, the state of Pennsylvania, for example, has long required every high school student to take at least one course in environmental studies, and other states are now following suit. Many states provide outdoor classroom experiences in parks, in coastal areas, or in other nature settings, and a growing number of young students across the country are pursuing serious environmental projects. However, well-trained teachers and appropriate instructional material are in short supply in many regions of the country, despite specialized teacher training programs that have been under way for many years. Unfortunately, the financial resources devoted to preparing the youth of the nation to participate in environmental programs have been grossly inadequate.

Let us look to the day when all high school graduates know that an estuary is a confluence of freshwater and saltwater bodies and that estuaries are vital marine habitats. Let us hope that bearers of high school diplomas will know that groundwater is more like an underground sponge than an underground river and that radon is formed during the radioactive decay of uranium in naturally occurring rock formations.

Many faculty members of our universities and colleges have an intense interest in important environmental issues and are eager to present courses, developing their own texts when necessary. Some higher educational institutions take great pride in their well-established environmental studies programs which combine mixes of science and the liberal arts. Other institutions offer environmental "familiarization" programs which we hope will not degenerate into simply easy courses for fulfilling mandatory science requirements. To participate effectively in the public debates on environmental issues, college graduates should understand that a standard deviation is a measure of uncertainty, and they should know how to find environmental regulations in the *Federal Register* which lines the shelves of libraries throughout the country.[1]

Formal education programs in the environmental field should stress the rigor of science and the centrality of ethics in environmental decisions. Many students are not equipped to deal with complex chemical formulas or intricate physical reactions. However, they all should appreciate the importance of care and precision in physics and chemistry which differentiate useful science from voodoo science. They should recognize the significance of encasing scientific conclusions within statements of certainty or uncertainty. While environmental ethics is still an evolving field, one moral precept is crystal clear. The environmental ethic demands a respect for the rights of future generations to continue to enjoy life on this planet.

The media will continue to dominate educational processes outside the classroom, and environmental disasters will always be featured on television and in the press. Environmental successes are simply less newsworthy. Who wants to hear about nontoxic coolants being used in new transformers when firemen are coping with the ecological threats of PCBs leaking from old transformers? Perhaps in the near term this bias toward sensationalism in the education of Americans is acceptable given the historical neglect of the environment and the need for more corrective actions now than in the past. However, in the longer term, more balanced reporting of environmental successes as well as problems will surely be needed.

* * *

Public support for governmental decisions is essential if national and local environmental policies are to achieve their aims. Public support cannot be commanded, nor can it be purchased. Support cannot be expected from an uninformed public nor from a public which has been excluded from the debates leading up to the decisions. The public wants to be heard, and indeed demands to be heard.

During the past decade the public trust of government has declined, and perhaps rightfully so. But public opposition to any proposal of government which is new can be self-defeating. Unfortunately, the public's Pavlovian response to government initiatives too often is, "Another cockamamy idea from Washington." And public enthusiasm for established environmental policies is often lethargic at best.

The EPA is slowly emerging from the depths of public mistrust where it has been mired following the outrageous behavior of some of its leaders in the early 1980s. Still, suspicions about the agency's actions abound in Washington and in other parts of the country. In particular, the EPA must become more sensitive to the importance of quantitative measures of environmental progress. The agency takes great pride in boasting that the levels of lead pollutants in the air have declined by 90%. Yet it pushes aside widespread concerns that only one-tenth of 1% of the 30,000 sites, which, in part, have been characterized as hazardous, have been completely cleaned up and that more than a decade is required to clean up a site. Also, the large number of cities which are out of compliance with air pollution standards after several decades of environmental controls is very disturbing to environmental constituencies, as are the tons of air toxic discharges every year and the hundreds of commercial chemicals which were identified in the 1970s for testing but have yet to be sent to the laboratory.

Of course the EPA is not the only federal agency which riles the anger of environmentalists. We have discussed the environmental liabilities at the nuclear weapons facilities of the Department of Energy. Many Army, Navy, and Air Force facilities throughout the country also have major environmental problems.[2] In addition, chemical contamination can be found at the wood and metal workshops of the Bureau of Prisons, in laboratories confiscated in federal drug raids, in silos filled with toxic fumigants owned by the government's Commodity Credit Corporation, and near abandoned mines on federal lands. No one can begin to estimate the environmental problems which will be encountered in the seizures of the assets of failed savings and loan associations.

The EPA and other agencies need to become more proactive in convincing the Congress and the public that they are putting their houses in order and that they are going beyond the requirements of environmental laws. If the number of cleaned up sites or facilities and the size of laboratory testing programs are not true indicators of progress, then the agencies should develop and articulate other easily understood indicators. Clearly, the government must constantly show its environmental commitment by both words and deeds if the public is to support the national effort.

Views of experts diverge on the details of almost every significant environmental proposal, old or new, and this divergence has to be accepted as a given. As indicated, of greater importance is a national commitment to stronger environmental protection and a public perception that public officials are living up to that commitment. Details of proposals will then be worked out with pain for a few but satisfaction for many. Arguing that regulations must accommodate compromises, many EPA officials smugly say, "If everyone is a little unhappy, we must have made the right decision." However, greater care is needed to ensure that such compromises do not erode the national commitment.

The United States sorely lacks charismatic leaders to rally the environmental instincts of a nation plagued by eco-risks. Such leaders could help bridge the gap between the suspicion of the public and the efforts of the government. A Jacques Costeau, a Carl Sagan, or a Ralph Nader for the environment simply has not emerged during the past 20 years. The administrators of the EPA are viewed as temporary captains of the ship. I hope that during the 1990s new leaders will step forth who can stir the emotions of the general public and command the attention of the government while promoting the ideas set forth by John Muir earlier in the century.

* * *

For 45 years we survived the cold war that divided East and West. As we prepare to enter the next century, we could be on the threshold of a hot war between nature and humanity. Environmentalists have joined the diplomats, generals, and bankers as full-fledged members of the national security establishments of most countries. Indeed, ecological security is rapidly becoming as important as military or economic security. All nations have been fighting local skirmishes to preserve their cities and towns, their rivers and beaches, and their forests and parks. Now we need broader alliances to combat the environmental mercenaries of global warming and ocean pollution. Since only humanity can be the winner or loser in environmental battles, all nations should be on the same side.

The United States should be among the leaders and not the followers in this international struggle confronting all nations. But we will be able to lead only if we have our own problems under control. The United States is the world's largest polluter and needs to reduce its own emissions if, for example, it is to encourage preservation of forests in the Amazon to help absorb these emissions.

The United States is the world's richest nation. If it is to lead, it needs to reorder budget priorities. Is one B-2 stealth bomber really twice as important as the annual research budget of the EPA? Should the United States continue to be among the stingiest of all donors of foreign aid? Can the states really be expected to shoulder the increasing economic burden of pollution control without greater help from Washington?

* * *

We can avoid an environmental apocalypse, but we don't have much time. According to the Administrator of the EPA, William Reilly, ". . . the United States does not now face an environmental crisis. Progress continues in abating some types of pollution problems in some places, and in the short haul no impending disasters can be predicted from a failure to address any of the lengthy list of environmental issues. Looking ahead, however, is a set of complex, diffuse, long-term environmental problems portending immense consequences for the economic well being and security of nations throughout the world, including our own . . . ."[3]

U.S. environmental policies during the 1990s, together with policies of other countries, will determine whether the next century is a time of prosperity or simply a time of survival. In its "Good Sense Formula for the 1990s," *Newsweek* wisely advocates:

> Stop splitting hairs: The most effective environmental standards are based on what's technologically feasible, not on arcane estimates of potential health hazards.
>
> Arrest the NIMBY (Not in My Back Yard) patrols: By blocking construction of new waste facilities, they keep bad old ones in operation.
>
> Regulate farms, not just factories: Only 9 percent of the pollutants flowing into America's streams come from industry. Sixty-five percent come from nonpoint sources.

> Save the swamps: Wetlands and mundane woods may lack the sex appeal of national parks and wildlife preserves but their ecological significance is greater.[4]

As to reasonable environmental targets for the United States, by the year 2000 all passenger automobiles should achieve 40 miles per gallon using cleaner fuels. Discharges of toxic pollutants into the air and water should be cut in half. With few exceptions our cities should be in compliance with ozone standards, and 95% of rivers and streams should be ecologically alive. Further degradation of groundwater should be capped, and chemicals used for farming should be reduced by more than 50%.

Within manufacturing industries, technologies which minimize waste and facilitate recycling should replace scrubbers and filters as the principal means of attacking pollution. Indeed, efficiency of plant operations is synonymous with pollution prevention. Leaking valves, wasted electricity, and discarded but valuable metals and organic chemical residues don't make sense—economically or environmentally.

While economic disparities among countries will persist, all countries can contribute to reducing the global pollution burden, to preserving the genetic richness of the flora and fauna, and to tempering the pressures on the natural resource base. By the year 2000 the United States should have shifted considerable financial resources from the Pentagon to foreign aid programs for resource conservation and pollution control. These activities should displace the Strategic Defense Initiative and mobile missiles in the national security debates in Washington. We must respond to new challenges to our security with our pocketbooks. At the same time, international cooperation in attacking environmental problems can build trust and confidence among countries which have been sorely lacking during the second half of the 20th century.

*   *   *

In sum, Americans have no choice but to pay now or pay later for their long-term survival. Actions or inactions during the 1990s will determine the costs during the next century. The environmental debt is accumulating, and the price tag for healing America's chemical wounds increases every year.

End Notes

Preface

1. For recent discussions of environmental trends, see "Environmental Progress and Challenges: EPA's Update," EPA-230-07-88-033, EPA, August 1988; "State of the Environment: A View toward the Nineties," The Conservation Foundation, Washington, D.C., 1987; and "Environmental Trends," Council on Environmental Quality, Government Printing Office, March 1990.
2. Weisskop, Michael, "Hypersensitivity to Chemicals Called Rising Health Problem," *Washington Post,* February 10, 1990, page 2.
3. Easterbrook, George, "Cleaning Up," *Newsweek,* July 24, 1989, page 27.

Chapter 1

1. Carson, Rachel, *Silent Spring,* Houghton Mifflin Company, Boston, 1962.
2. Whitaker, John C., "Earth Day Recollections: What It Was Like When the Movement Took Off," *EPA Journal,* July/August 1988, page 16.
3. For a description of early EPA activities to curb pollution, see Quarles, John, *Cleaning Up the Environment: An Insider's View of the Environmental Protection Agency,* Houghton Mifflin Company, Boston, 1976, page 12.
4. "Toxic Substances," prepared by the Council on Environmental Quality, April 1971. For elaboration of some of the concerns, see "Chemicals and Health," Report of the Panel on Chemicals and Health of the President's Science Advisory Committee, National Science Foundation, September, 1973.
5. Toxic Substances Control Act of 1976, Section 2 (c).
6. Federal Water Pollution Control Act (as amended) of 1972, Section 307 (a); and Clean Air Act of 1972, Section 112.
7. See, for example, Epstein, Samuel, *The Politics of Cancer,* Sierra Books, 1978. For a recent discussion of the causes of cancer, see Gough, Michael, "Estimating Cancer Mortality," *Environmental Science and Technology,* August 1989, pages 925–930.
8. Wade, Nicholas, "Control of Toxic Substances: An Idea Whose Time Has Nearly Come," *Science,* February 13, 1976, pages 541–545.
9. For overviews of EPA activities leading up to passage of the Toxic Substances Control Act, see "A Framework for the Control of Toxic Substances," Office of Toxic Substances, EPA, April 1975; and "Selected Aspects of the Control of Toxic Substances," Office of Toxic Substances, EPA, May 1976.

10. See, for example, "Toxic Substances: EPA's Chemical Testing Program Has Made Little Progress," U.S. General Accounting Office, GAO/RCED-90-112, April 1990.
11. "Settlement Agreement," Civil Actions 2153-73, 75-0172, 75-1698, and 75-1267, in the United States District Court for the District of Columbia, June 7, 1976.
12. Marx, Jean L., "Drinking Water: Another Source of Carcinogens," *Science,* November 29, 1974, pages 809–810.
13. For early EPA efforts to control toxic chemicals, see "Summary Tabulation of Selected EPA Activities concerning Toxic Chemicals," Office of Toxic Substances, EPA, April 1976.
14. "Environmental Group Ranks Toxic Pollutants," *Washington Post,* August 11, 1989, page A10.
15. "Taking Inventory of 7 Billion Toxic Pounds," *USA Today,* August 1, 1989, pages 6A–7A.

Chapter 2

1. Long, Janice R., and David J. Hanson, "Dioxin Issues Focus on Three Major Controversies in US," *Chemical and Engineering News,* June 6, 1983, pages 23–36.
2. Gladwell, Malcolm, "Scientists Temper Views on Cancer-Causing Potential of Dioxin," *Washington Post,* May 31, 1990, page A3.
3. Schweitzer, Glenn E., "Toxic Substances: Legislation, Goals, and Case Studies," *A Framework for the Control of Toxic Substances,* Office of Toxic Substances, EPA, April 1975, page 4.
4. Berry, D. Kent, "Air Toxics," *Environmental Science and Technology,* Volume 20, Number 7, 1986, pages 647–651; also, see "National Emission Standards for Hazardous Air Pollutants," EPA, 40 CFR Part 61, September 14, 1989, for a discussion of technical considerations in setting a benzene standard as an example of current regulatory approaches.
5. Ember, Lois R., "President's Clean Air Bill Gets Mixed Reviews," *Chemical and Engineering News,* August 7, 1989, page 26.
6. *Environmental Monitoring at Love Canal,* Volume 1, EPA 600/4-82-030A, EPA, May 1982, pages 21–22.
7. Abelson, Philip H., "The Asbestos Removal Fiasco," *Science,* May 2, 1990, page 1017.
8. "Lead Contamination Control Act," EPA 570/9-89-AAA, EPA, July 1989.

Chapter 3

1. Technical documentation about the vinyl chloride case is included in: "Preliminary Assessment of the Environmental Problems Associated with Vinyl Chloride and Polyvinyl Chloride (with Appendices)," EPA, September 1974.
2. A good discussion of the pervasiveness of PCBs in the mid-1970s is included in Maugh, Thomas H., "Chemical Pollutants: Polychlorinated Biphenyls Still a Threat," *Science,* December 19, 1975, page 1189.

3. For a recent discussion of chemical carcinogens, see "Special Report: A Brief Review of Chemical Carcinogenesis," The Texas Institute for Advancement of Chemical Technology, Special Report 1, 1989, College Station, Texas.
4. "Water Program: Benzidine; Proposed Toxic Pollutant Effluent Standards," *Federal Register,* Part V, June 30, 1976. This notice reviews development of the quantitative standard after it was first proposed in the *Federal Register* on December 27, 1973.
5. "Auto Pollution Health Costs Calculated," *Washington Post,* January 21, 1990, page A12.
6. *Valuing Health Risks, Costs, and Benefits for Environmental Decision Making,* National Research Council, National Academy Press, 1990, page 207.
7. A good summary of early efforts to standardize risk assessment methodologies for carcinogens is included in Rushevsky, Mark E., *Making Cancer Policy,* State University of New York Press, 1986.
8. "Scientific Bases for Identification of Potential Carcinogens and Estimation of Risks," Report of the Interagency Regulatory Liaison Group, Work Group on Risk Assessment, 1979, page 1.
9. Ames, Bruce N., "Identifying Environmental Chemicals Causing Mutations and Cancer," *Science,* May 11, 1979, pages 587–593.
10. For more detailed discussions of ecological risk assessment, see Messer, Jay J., "Keeping a Closer Watch on Ecological Risks," *EPA Journal,* May/June 1989, pages 34–36; and Bascietto, John, Dexter Hinckley, James Plafkin, and Michael Stinck, "Ecotoxicity and Ecological Risk Assessment, Regulatory Applications at EPA," *Environmental Science and Technology,* Volume 24, Number 12, 1990, pages 10–15; and "Review of Ecological Risk Assessment Methods," EPA AX 8907-0100, EPA, 1988.
11. Wingard, Laura, "Lake Project Continues To Be Studied," Las Vegas *Review Journal,* November 15, 1987, page 7B; Beall, Christopher, "Lake Lovers Return Something to Mead," Las Vegas *Review Journal,* May 31, 1987, page 7B.
12. EPA has widely publicized its Environmental Monitoring and Assessment Program which begins pilot testing in 1990 directed to detecting changes in indicators of ecological conditions—terrestrial, freshwater, and coastal ecosystems. See, for example, Bromberg, Steven M., "Identifying Ecological Indictors: An Environmental Monitoring and Assessment Program," *Journal of the Air and Waste Management Association,* July 1990, pages 976–978.
13. An interesting discussion of uncertainty is included in Goldstein, Bernard D., "The Problem with the Margin of Safety: Toward the Concept of Protection," *Risk Analysis,* Volume 10, Number 1, 1990. Also, see "Health Hazards: An Imperfect Science," *Technology Review,* May/June 1986, pages 60–75.
14. National Academy of Sciences Symposium on Risk Communications, September 11, 1989.
15. Ruckleshaus, William D., "Risk, Science, and Democracy," *Issues in Science and Technology,* Spring 1985, pages 22–26.
16. See, for example, *Decision Making for Regulating Chemicals in the Environment,* National Research Council, National Academy Press, 1975; *Risk Assessment in the Federal Government,* National Research Council, National Academy Press, 1983; and "Risk Assessment and Management," EPA 600/9-85-002, EPA, December 1984. For more specificity, see Paustenbach, Dennis J. (editor), *The Risk Assessment of Environmental*

and *Human Health Hazards: A Textbook of Case Studies,* John Wiley and Sons, New York, 1989.
17. "Judge Bazelon's Brilliant Address on Role of Courts in Health Improvement," *Occupational Health and Safety Letter,* Washington, October 22, 1980, pages 3–5.
18. See, for example, Fiksel, Joseph, "Victim Compensation," *Environmental Science and Technology,* Volume 20, Number 5, 1986, page 425.
19. Schweitzer, Glenn E., "Relevance of Radiation Compensation Litigation to Compensation for Toxic Exposures," *Environmental Monitoring and Assessment,* Volume 8, 1987, pages 1–10.

Chapter 4

1. Greenberg, Michael R., David B. Sachsman, Peter M. Sandman, and Kandice L. Salomone, "Network Evening News Coverage of Environmental Risk," *Risk Analysis,* Volume 9, Number 1, 1989, page 119.
2. *Chemical Risks: Fears, Facts, and the Media,* The Media Institute, Washington, D.C., 1985, page xii.
3. *Alternative Agriculture,* National Research Council, National Academy Press, Washington, D.C., 1989.
4. Silva, Mark, "New Migraine for Motorists: Mandatory Testing of Exhausts," *The Miami Herald,* September 11, 1989, pages 1A and 8A.
5. Krimskey, Sheldon, and Alonzo Plough, *Environmental Hazard: Communicating Risks as a Social Process,* as reported in *Chemical and Engineering News,* January 30, 1989.
6. Covello, Vincent T., "Communicating Right-to-Know Information on Chemical Risks," *Environmental Science and Technology,* Volume 23, Number 12, 1989, pages 1444–1450.
7. "Chemical Risk Communication: Preparing for Community Interest in Chemical Release Data," American Chemical Society, 1988. "Seven Cardinal Rules of Risk Communication," OPA-87-020, EPA, April 1988. See also "Exploring Environmental Risk," Office of Toxic Substances, EPA, November 1986; and *Improving Risk Communication,* National Research Council, National Academy Press, 1989.
8. "Agency Operating Guidance, FY 1985–1986," Office of the Administrator, EPA, February 1984, page 2.
9. Cronin, Patti, "Mediation: How It Worked in East Troy, Wisconsin," *EPA Journal,* March/April 1987, pages 46–47.
10. "Unfinished Business: A Comparative Assessment of Environmental Problems," Volume 1, Overview, EPA, February 1987.

Chapter 5

1. Reilly, William K., "A Management Report of the Superfund Program," EPA, 1989. Also, letter from EPA Deputy Assistant Administrator Mary Gade, October 26, 1990.
2. Bush, George, Speech to State Governors Conference, June 12, 1989.

3. *Coming Clean: Superfund Problems Can Be Solved,* Office of Technology Assessment, U.S. Congress, October 1989.
4. Interviews at the EPA, July 1990.
5. *Understanding Superfund: A Progress Report,* R-3838-ICJ, Rand Publications, Santa Monica, California, September 1989.
6. "Superfund: Looking Back, Looking Ahead," Reprinted from *EPA Journal,* OPA 87-007, EPA, December 1987, page 1.
7. See "The New RCRA, A Fact Book," EPA/530-SW-85-035, EPA, 1985.
8. Lewis, Jack, "What's in the Solid Waste Stream?" *EPA Journal,* March/April 1989, pages 15–17.
9. For a review of the history leading to the Yucca Mountain project, see Carter, Luther J., "Nuclear Imperatives and Public Trust: Dealing with Radioactive Waste," *Issues in Science and Technology,* Winter 1987, pages 46–62. Also, see letter from Secretary James B. Watkins to Senator J. Bennett Johnston of March 1, 1990, released by Department of Energy on March 6, 1990.
10. See "The DOE Position on the MRS Facility," Office of Civilian Radioactive Waste Management, Department of Energy, DOE/RW 0239, June 1989.
11. Frederick, Sherman R., "Nukes and Us," Las Vegas *Review Journal,* March 18, 1990, Editorial Page.
12. Thompson, Lloyd H., "Test Site Solutions," Las Vegas *Review Journal,* March 24, 1990, Editorial Page.
13. See, for example, "Trouble at Rocky Flats," *Newsweek,* August 14, 1989, page 19.
14. Glenn, John, "National Security: More than Just Weapons Production," *Issues in Science and Technology,* Summer 1985, pages 27–28.
15. For a summary of current EPA groundwater protection activities, see "Progress in Ground-Water Protection and Restoration," EPA 440/6-90-001, February 1990. Also, for an overview of the national groundwater situation, see "Improved Protection of Water Resources from Long-Term and Cumulative Pollution: Prevention of Ground-Water Contamination in the United States," Prepared for the Organization of Economic Co-operation and Development, EPA, April 1987.

Chapter 6

1. "The Maryland Initiative: Lesson for the Nation," *EPA Journal,* September/October 1989, pages 29–30.
2. "E Pluribus, Plures," *Newsweek,* November 13, 1989, page 71.
3. Personal communication with staff of the National Governor's Association, September 1989.
4. *Goals for State–Federal Action, 1989–1990,* National Conference of State Legislatures, 1989.
5. "States Bear Growing Share of Environmental Cleanup Cost," *Chemical and Engineering News,* September 11, 1990, pages 19–20.
6. Kean, Tom, "Dealing with Air Toxics," *Issues in Science and Technology,* Summer 1986, page 22.

7. "ECRA Report—FY 1989," New Jersey Department of Environmental Protection, 1989.
8. Breckstrom, Linda, "State Environmental Initiative Would Create Elected Advocate," San Francisco *Herald Examiner,* October 11, 1989, pages A-3 and A-8.
9. For a discussion of this definition and many related runoff issues, see Thompson, Paul, *Poison Runoff, A Guide to State and Local Control of Nonpoint Source Water Pollution,* Natural Resource Defense Council, Washington, April 1989.
10. One excellent example of state efforts to minimize adverse impacts of timber operations is "Silviculture, Best Management Practices Manual," Florida Department of Agriculture and Consumer Services, Division of Forestry, May 1990.
11. O'Connor, Charles A., and Donna G. Diamond, "Current Development under Proposition 65," unpublished manuscript provided to the American Chemical Society, June 1988.
12. Personal communication with the staff of the National Governor's Association, January 15, 1990.
13. "Summary of State Commissioners' Meeting on EPA Proposed Strategy on Agricultural Chemicals in Groundwater," Office of Pesticides and Toxic Substances, EPA, August 1988, pages 10–12. See also "Florida Groundwater Protection Task Force, Annual Report 1988–1989," Florida Department of Environmental Regulation, October 1989.
14. State Commissioners' Meeting, page 7.
15. See, for example, Workman, Bill, "60-Square Mile Medfly Quarantine Area," and Wildermath, John, "Medfly Spraying Called Safe for People, Pets," *San Francisco Chronicle,* September 7, 1989, page A4.
16. "Toxic Waste Program Great," *Marin Independent Journal,* September 30, 1989, page A6.
17. *A Matter of Chance, A Matter of Choice, Living with Environmental Risks in Wisconsin,* University of Wisconsin–Madison, 1989.
18. Romer, Roy, "An Elected Official," *EPA Journal,* November/December 1988, page 15.
19. *1988 Biennial Report to the Legislature,* Minnesota Pollution Control Agency, December 31, 1988, page 34.

Chapter 7

1. *Code of Federal Regulations,* 40 CFR 262.23 and 40 CFR 262.41 (a)(6) and (7).
2. *CMA Waste Minimization Resource Manual,* Chemical Manufacturers Association, Washington, D.C., 1989, page 3-3.
3. Higgins, Thomas E., *Hazardous Waste Minimization Handbook,* Lewis Publishers, Chelsea, Michigan, 1989, page xvii.
4. *Title III: One Year Later, Plant Manager Interviews,* Chemical Manufacturers Association, June 1989.
5. *Did You Know?* Dow Chemical Company, 330-00527-889, undated, available 1989.
6. *National Chemical Response and Information Center,* Chemtrac, Chemnet, Chemical Referral Center, Training Programs, Chemical Manufacturers Association, undated, available 1989.

7. Ableson, Philip, "Regulation of the Chemical Industry," *Science,* November 3, 1978, page 473.
8. "Code Grades Corporate Ecology," *Washington Post,* September 16, 1989, page D19.
9. *Responsible Care,* Chemical Manufacturers Association, undated, available 1989.

Chapter 8

1. An interesting history of stratospheric ozone depletion is presented in Broeder, Paul, "Annals of Chemistry, in the Face of Doubt," *The New Yorker,* June 9, 1986, pages 70–87.
2. Benedick, Richard Elliot, "Ozone Diplomacy," *Issues in Science and Technology,* Fall 1989, pages 43–50.
3. "Ozone Depletion Accord,"*Chemical and Engineering News,* July 9, 1990, page 6.
4. "CFC Substitutes, Candidates Pass Early Toxicity Tests," *Chemical and Engineering News,* October 9, 1989, pages 4–5; Manzer, L. E., "The CFC-Ozone Issue: Progress in the Development of Alternatives to CFCs," *Science,* July 6, 1990, pages 31–35.
5. See, for example, Wilson, R. T., M. A. Geller, R. S. Stolarski, and R. F. Hampson, *Present State of Knowledge of the Upper Atmosphere, An Assessment Report,* Reference Publication 1162, NASA, May 1986.
6. Rind, David, "A Character Sketch of Greenhouse," *EPA Journal,* January/February 1989, page 7.
7. White, Robert M., "Uncertainty and the Greenhouse Effect," public statement released by the National Academy Op-Ed Service, National Academy of Sciences, Washington, D.C., August 27, 1989. Also, see White, Robert M., "The Great Climate Debate," *Scientific American,* July 1990, pages 36–43.
8. "Policy Makers Summaries," Reports of Working Groups I, II, and III of the Intergovernmental Panel on Climate Change, World Meteorological Organization and United Nations Environmental Program, June 1990.
9. Brown, Lester R., *et al., State of the World 1990,* Worldwatch Institute, Norton and Company, New York, 1990, page 176.
10. " '89: Policy Implications of the GRI Baseline Projections of US Energy Supply and Demand to 2010," Gas Research Institute, undated but distributed in 1989, page 1.
11. Schweitzer, Glenn E., "Toxic Chemicals: Steps toward their Evaluation and Control," *Environmental Protection, the International Dimension,* Allanheld, Osmun, Publishers, Montclair, N.J., 1983, pages 29–31.
12. Mathews, Jessica Tuchman, "Redefining Security," *Foreign Affairs,* Spring 1989, page 162.

Chapter 9

1. Text provided by the office of Texas Commissioner of Agriculture, December, 1989.
2. *Alternative Agriculture,* National Research Council, National Academy Press, 1989.

3. See, for example, Schneiderman, Howard A., and Will D. Carpenter, "Planetary Patriotism: Sustainable Agriculture for the Future," *Environmental Science and Technology,* April 1990, pages 466–473.
4. *Introduction of Recombinant DNA-Engineered Organisms into the Environment, Key Issues,* National Research Council, National Academy Press, 1987.

Chapter 10

1. See, for example, Weis, Judith S., "The Status of Undergraduate Programs in Environmental Sciences," *Environmental Science and Technology,* August 1990, pages 1116–1121.
2. The report for 1987 and 1988 of the White House Council on Environmental Quality presented an unwarranted optimistic view of the environmental situation on the lands and at the installations of the Department of Defense, for example. See "Environmental Quality," Council on Environmental Quality, Executive Office of the President, undated but released in late 1989, pages 149–181. Shortly after its publication, *Newsweek* provided an insightful revelation of the very severe environmental problems at some of the worse sites of the Department of Defense. See Turque, Bill, and John McCormick, "The Military's Toxic Legacy," *Newsweek,* August 6, 1990, pages 20–23.
3. *State of the Environment: A View toward the Nineties,* A Report of the Conservation Foundation, Washington, D.C., 1987, page xxxix.
4. Easterbrook, Gregg, "Cleaning Up," *Newsweek,* July 24, 1989, page 29.

Index

Acid rain
ecological impact and, 29–30
international cooperation and, 236
remote sensing and, 256
Aerial detection, pollutants, 253–257
Aerosol sprays, ozone depletion and, 216–217
Agency for International Development (AID), programs of, 2–4
Agent Orange, 40, 41, 44
Agriculture
alternative approaches, 261–263
biotechnology, 264–267
chemical runoff and, 173
groundwater contamination and, 161, 180
ozone depletion and, 215–216
pesticides and, 7–8
Air pollution
acid rain and, 236
environmental awakening and, 4, 5
EPA and, 25–26
fossil fuels and, 227–229
incinceration and, 189–191
international cooperation and, 237–238
legislation and, 8, 9
national standard setting and, 44–48
persistence of, 272
politics and, 226
public health concerns and, 44
radioactivity and, 46
remote sensing of, 253
Aldrin, water quality criteria, 24
American Chemical Society, xii, 106
Animal studies, risk assessment and, 77–78, 80
Arsenic, air pollution standards, 46
Asbestos
ecological hazards of, 14
emissions of, 26
Johns Manville Co. and, 98
air pollution, standards for, 45

Baker, James, 213
Bazelon, David L., 95–96
Benzene
air pollution standards for, 46
emissions of, 26
groundwater pollution by, 250
Beryllium
emissions of, 26
standards for, 45
Bhopal, India, tragedy, 97, 197, 231
Biotechnology, 264–267
Birth defects, public health and, 29
Breast milk, PCB contamination of, 72–76
Burden of proof, EPA and, 11–12, 13
Bureau of Reclamation, projects of, 3
Bush, George, 156
air pollution and, 47
Clean Water Act and, 170
EPA and, 139
hazardous waste disposal and, 135
population programs of, 235
Butadiene, regulation of, 46

Cadmium
levels of, 8
regulation of, 46
water pollution by, 24
Canada, 236, 241
Cancer. *See* Carcinogens; Skin cancer
Carbon dioxide, greenhouse gases and, 221–227

Carbon tetrachloride, regulation of, 46
Carcinogens
  animal studies and, 77–78, 80
  asbestos, 45
  environmental awakening and, 7–10
  nutrition and, 83
  politics and, 15
  public health and, 29
  risk assessment by quantification and, 78–79, 82–83
  solvents, 193–194
  toxic chemical categorization and, 50–51
  vinyl chloride, 65–72
  water pollution and, 25
Carson, Rachel, 5, 28, 249
Carter, Jimmy, 21
Catalytic converters, 260
Centers for Disease Control, 60; *see also* Love Canal
Chemical industry
  environmental consciousness, 196–200
  international cooperation and, 238–239
  public relations and, 205
  transportation accidents and, 206–208
Chemical Manufacturers Association (Manufacturing Chemists Association), 17, 211
Chemical runoff, state agencies, interest in, 173–176
Chemistry, definition of, 1
CHEMNET, 208
CHEMTREC, 208
Chernobyl, U.S.S.R., disaster, 230–231
Chesapeake Bay, 15, 28, 165–168
China, coal in, 273
Chlorofluorocarbons (CFC's)
  greenhouse gases and, 225
  limitations on, 21
  Montreal Protocol and, 219
  ozone depletion and, 19, 214–221
Chloroform, regulation of, 46
Chromium, regulation of, 46
Clean Air Act, 26, 28, 47, 48
Clean Water Act, 28, 170, 173
Climate
  greenhouse gases and, 221–227
  ozone layer and, 213–221
Communication: *see* Risk communication
Consent Decree, 23–25; *see also* National Resources Defense Council
Conservation measures, x
Constitution, U.S. Amendment X, 165
Corps of Engineers, projects of, 3
Cost/benefit analysis
  judiciary and, 94–98
  legal requirements, 11
  PCBs, 72–76
  policy implications of, 91–94
  risk analysis by quantification and, 76–84
  vinyl chloride and, 65–72
Costs
  air pollution, 79
  environmental protection and, xi
  losses due to ecological damage, 88
  See also Financial concerns
Council on Environmental Quality, establishment of, 8
Courts: *see* Judiciary; Law
Crisis, definition of, 35

Dams, water quality and, 2
DDT, ix, 265
Department of Health and Human Services, 74
Developing countries
  energy policy and, 234
  environmentalism and, 2–4, 247
  pollution reduction and, 234
  population growth in, 273
  tax policies and, 242
Dichloromethane, characteristics of, 32
Dieldren, water quality criteria, 24
Diet. *See* Nutrition
Dioxin, ix, 40–44
  concentration levels of, 53
  Times Beach contamination, 41–43
  uncertainty and, 91
DNA biotechnology, 264–265
Dole, Robert, 65
Dow Chemical Company, 36–37

Earth Day, 5
Ecological impact, 213–247
  environmental awakening and, 28–30

Ecological impact (*cont.*)
fossil fuels and, 227–229
greenhouse gases, 221–227
international cooperation and, 236–240
international negotiations and, 240–244
nuclear energy and, 229–236
ozone depletion and, 213–221
risk assessment and, 86–89
science, scientists, and, 246
security aspects and, 244–247
Eco-diplomacy, 236
Ecology, definition of, 88
Economic factors. *See* Financial concerns
Education: *see* Public education
Energy production
fossil fuels, 227–229
greenhouse gases reduction, 257–261
nuclear energy, x, 229–236
Environmental accidents
Bhopal, India, 97, 197, 231
Chernobyl, 230
Three-Mile Island, 97, 110–111, 114, 149, 230
Valdez, Alaska, 5, 197
Environmental advocate, 172
Environmental awakening, 1–34
alternative control routes and, 23–28
American spirit and, 4–7
carcinogens and, 7–10
chemical industry and, 196–200
dimensions of, 30–33
ecological impact and, 28–30
foreign aid and, 1–4
future prospects, 33–34
politics and, 16–18
regulatory administrators and, 18–23
Toxic Substances Control Act of 1976, 10–16
water pollution, 5
Environmental Impact Statements, 2, 3
Environmental organizations, EPA and, 23
EPA: *see* U.S. Environmental Protection Agency (EPA)
Epidemiology, 51–52, 80
Ethylene dibromide, hazards from, 14
Ethylene dichloride, regulation of, 46
Exposure time, risk analysis by quantification, 81

Federal government decisions
public support for, 276–277
state governments and, 169, 187
See also EPA
Fiber-optic technology, groundwater pollution and, 249–252
Financial concerns
chlorofluorocarbons and, 214
developing nations and, 242
ecological impact and, 29–30
environmental awakening and, 6
EPA and, 21–22
greenhouse gases and, 226
hazardous waste disposal and, 27–28, 130
health care, 79
hexachlorobenzene and, 36, 38
industry and, 197
industry waste minimization efforts, 200–203
international cooperation and, 238
nuclear waste disposal and, 158
state agencies and, 168–172
Superfund and, 137–138
Finch, Bob, 5
Ford, Gerald, 16–17
Foreign aid, environmentalism and, 2–4
Formaldehyde, indoor pollution from, 30
Fossil fuels
China, 273
ecological focus and, 227–229
Freedom of Information Act, 14, 121
Freon, 19
Fuel switching, 259
Fuller, Buckminster, 189

Gasoline, lead in, 25–26
Gasoline storage tanks, hazardous waste disposal, 147–149
General Accounting Office, 16
Genetic damage, public health and, 29
Genetic engineering: *see* Biotechnology
Glenn, John, 158
Global concerns: *see* Ecological impact

Global warming: *see* Greenhouse gases
Goodrich company, vinyl chloride and, 66
Gorbachev, Mikhail, 236
Greenhouse gases
ecological impact of, 221–227
financial concerns and, 226
fossil fuels and, 227, 273
international cooperation and, 221, 225–226, 235
reduction of, 257–261
Groundwater
chemical runoff and, 173–174
hazardous waste disposal and, 144, 159–163
pollution and EPA, 249–252
state agencies and, 177, 180–182

Hardin, Clifford, 5
Hazardous level assessment
dioxin, 41–42
EPA and, 8
hazardous waste disposal and, 53
hexachlorobenzene, 38–39
risk assessment by quantification and, 76–84
testing problems and, 55–61
vinyl chloride, 66
See also Testing problems
Hazardous waste disposal, 129–163.
analytical problems of, 52–55
dioxin, 41
environmental awakening and, 26–27
financial concerns and, 27–28, 129–130
groundwater and, 153, 159–163
hexachlorobenzene, 36–37
incineration and, 190–193
industry minimization programs and, 200-203
international cooperation and, 243–244
municipal wastes, 145–146
nuclear waste, 149–158
Resource Conservation and Recovery Act, 143
safety issues in, 143–147
source of hazardous waste, 137
Superfund and, 135–143
surveys of needs for, 129–135
Hazardous waste disposal (*cont.*)
technology and, 24
underground storage tanks and, 147–149
U.S. Department of Energy and, 115, 152–158
See also Superfund
Heckert, Richard, 189
Hexachlorobenzene, 35–40
Hickel, Walter, 5
Hiroshima, Japan, 151
Hooker Chemical Company, 58; *see also* Love Canal
Huxley, Thomas, v
Hydropower, 258–259

Incineration
hazardous waste disposal by, 144–145
industry and, 190–193
Industry, 189–212
chemical industry, 196–200
chlorofluorocarbons and, 216
environmental consciousness, 196–200
future changes for, 208–212
hazardous waste, financial responsibility for, 139
minimization of waste by, 200–203
public relations and, 198, 203–205
risk assessment and, 194–195
safety training programs by, 210
Sheldahl Company example, 193–196
social engineering and, 268
3M Company example, 189–193
timing of data release, 120–121
transportation accidents and, 206–208
Insurance industry, protection by, 197
International cooperation
chemicals and, 238–239
developing nations and, 234
ecological impact and, 236–240
greenhouse gases, 221, 225, 226
levels of negotiation, 240–244
ozone depletion and, 217–219
pesticides and, 237
security aspects, 244–247
technology and, 238
waste discharges and, 241
water pollution and, 237–238, 241

Japan, 151, 225, 237–238
Jefferson, Thomas, 103
Johns Manville Company, 98
Judiciary
  EPA and, 94–98, 209
  industry responsibility and, 197–199
  risk assessment and, 94–98
  See also Law

Keller, Helen, 65
Kepone, water contamination and, 14–15

Lake Erie, environmental awakening and, 5
Lake Mead, ecology of, 87–88
Lake pollution, aerial/space detection of, 254–256
Landfills
  concentration levels, 54
  hazardous waste disposal and, 27, 144
  municipal waste and, 145–146
Laser technology
  aerial detection of pollution, 253–257
  groundwater pollution and, 249–252
Law
  air and water pollution, 8, 9
  cost/benefit analysis, 11
  ozone depletion and, 19
  risk assessment and, 94–98
  waste minimization efforts and, 201
  See also Judiciary
Lead
  levels of, 8
  regulation of, 63
  water pollution by, 24
Leaded gasoline, 25–26
Legislation: *see* Law
London Dumping Convention, 244
Love Canal, New York, 55–61, 111–114, 120–130

Malathion, insect infestation and, 182–183
Manufacturing Chemists Association (Chemical Manufacturers Association), 17, 211
Mathews, Jessica Tuchman, 244
Mayo, Bob, 6
McCracken, Paul, 6
Media
  PCBs and, 72–76
  print, 107–109
  public education and, 276
  public meetings and, 109–114
  PVC and, 104–105
  television, 103–107
Mediation, risk communication, 122–125
Mediterranean fruit fly, 182–183
Mercury
  assessment of toxicity of, 8
  emissions of, 26
  environmental awakening and, 5
  identification of, 53
  standards for, 45–46
  water pollution by, 24
Methylene chloride, emissions of, 194–196
Minimization of waste, industry efforts, 200–203
Mining industry, chemical runoff and, 175
Mixtures, risks from, 84–86
Monsanto Company, PCBs and, 10
Montreal Protocol, 219
Multispectral sensing technology, pollution detection by, 254
Municipal waste, hazardous waste disposal and, 145–146

Nagasaki, Japan, 151
National Academy of Sciences, studies of, 45, 103, 108, 261, 262, 266
National Aeronautics and Space Administration, 215
National Audubon Society, nuclear reactors and, 232
National Conference of State Legislatures, 170
National Environmental Policy Act of 1969, 1–2
National Science Foundation, 215
National security. *See* Security aspects
National Resources Defense Council, 23, 58
  Consent Decree and, 23–28
  EPA and, 124

National Resources Defense Council (*cont.*)
Love Canal and, 58
Toxic Substances Control Act and, 17
Nevada Test Site, 113–114, 151
Nixon, Richard M., 5–6, 10, 93–94
North Eastern Pharmaceutical and Chemical Company, dioxin and, 40
Nuclear bomb, radiation from, 151
Nuclear energy
energy production, x, 149–150, 229–336
greenhouse gases reduction, 257–261
safety considerations and, 150
Nuclear waste disposal, ix–x, 149–158
burial sites for, 152–156
nuclear power production and, 229–230
nuclear weapons plants and, 156–158
political factors in, 154–156
public meetings and, 109–114
sources of waste, 149–150
surface storage of, 153–154
technical considerations in, 151
U.S. Department of Energy and, 115
Nuclear weapons plants, nuclear waste disposal and, 156–158
Nursing mothers, PCBs and, 72–76
Nutrition, cancer and, 83

Office of Management and Budget, 11, 17, 76
Oil spills, environmental awakening and, 5
Organization for Economic Cooperation and Development, 239, 240
Ozone depletion
ecological impact and, 30, 213–221
international cooperation and, 217–219
regulation of, 19
skin cancer and, 216

PCBs: *see* Polychlorinated biphenyls (PCBs)
Perchloroethylene, 46
Pesticides
agriculture and, 4, 7–8
Bhopal tragedy, 97
biotechnology and, 264–267
environmental awakening and, 5, 28
Pesticides (*cont.*)
EPA and, 12–13
export of, 3–4
groundwater and, 180–182
international cooperation and, 237
Kepone, 14–15
public health and, 180
rebuttable presumption, 12
technological reduction in use of, 262–263
Politics
carcinogens and, 15
Chesapeake Bay and, 165–166
data release, 1209
environmental awakening and, 16–18, 20
EPA and, 12, 277–278
greenhouse gases and, 226
hazardous waste disposal and, 131, 135–136, 139, 142, 148
nuclear waste disposal and, 154–156
pesticides and, 7–8
press and, 107–109
public health and, 29
public meetings, 109–115
television and, 103–107
Toxic Substances Control Act of 1976 and, 15–16
Pollutant removal. *See* Hazardous waste disposal
Polychlorinated biphenyls (PCBs)
air pollution and, 44
biotechnology and, 265
breast milk and, 72–76
cost/benefit analysis, 72–76
environmental problems of, 9–10
regulation of, 19
Polyvinyl chloride (PVC)
breast milk and, 42–46
cancer and, 66–69
media and, 103–104
Population program, 235, 245
Premanufacturing notification concept, 13, 22
Public education
necessity of, 275–2765
state agencies and, 182–185

Public health concerns
  air pollution and, 44
  ecological impact contrasted, 28–29
  economic aspects, 37
  EPA standards for chemicals and, 46
  governmental overreaction to, 62
  nuclear waste disposal and, 151, 158
  PCBs and, 74–76
  pesticides and, 180
  politics and, 17–18
Public Health Service, 74
Public meetings
  openness and, 127
  risk communication, 109–115
Public policy
  energy conservation in developing countries and, 234
  future requirements and, 279–280
  risk assessment, 91–94
Public relations departments
  industry and, 198, 203–205
  data release and, 120–121

Radioactive air pollutants, 46
Radioactive hazards: *see* Nuclear waste disposal
Rain forest, development and, 2
Reagan, Ronald, 21, 131
Recombinant DNA biotechnology, 264–267
Redford, Robert, global warming interest of, 241
Regulatory administrators, environmental awakening and, 18–23
Reilly, William, 271, 279
Resource Conservation and Recovery Act, 143
Revolving door regulatory administrators, 18–23
Risk assessment, 65–101
  air pollution standards and, 44–47
  burden of proof and, 11–12
  ecological resources and, 86–89
  future risk reductions, 98–101
  individual chemicals and, 49–50
  industry and, 194–195
  judiciary and, 94–98
Risk assessment (*cont.*)
  mixtures of chemicals, 84–86
  PCBs, 72–76
  pesticides and, 12
  policy implications of, 91–94
  quantification and, 76–84
  rebuttable presumption and, 12
  Times Beach and, 91
  uncertainty and, 89–91
  vinyl chloride, 67–72
Risk communication, 103–125
  conventional wisdom as to, 115–118
  mediation versus confrontation, 122–125
  openness and, 125–128
  press and, 107–109
  public meetings and, 109–115
  television and, 103–107
  timing of data release, 118–122
Rogers, Bill, 5
Roosevelt, Theodore, 271
Ruckleshaus, William, 90
Russell, Bertrand, 65

Safe Drinking Water Act (U.S.), 25
Safe Drinking Water and Toxic Enforcement Act (California), 28, 177–178
Safety training programs, industry, 210
Sampling. *See* Hazardous level assessment; Testing problems
Satellite detection, pollutants, 253–257
Security aspects, ecological focus, 244–247
Sheldahl Company, 193–196
Shevarnadze, Eduard, 213
Sierra Club, 15–16, 17
Skin cancer, ozone depletion and, 216
Social engineering, technology and, 267–270
Soil pollution, dioxin, 41–43
Solvents, 193–194
Soviet Union. *See* Union of Soviet Socialist Republics
Stans, Maurice, 6
State agencies, 165–187
  chemical runoff impacts and, 173–176
  Chesapeake Bay and, 165–168
  education and, 182–183

State agencies (*cont.*)
environmental advocate, 172
EPA and, 170
financial considerations and, 168–172
future prospects and, 185–187
groundwater protection and, 180–182
public education and, 182–185
toxics and, 176–180
water pollution and, 24
Sulfuric acid, characteristics of, 32
Superfund. *See also* Hazardous waste disposal
Bush, George, and, 135
creation of, 22
engineering service companies and, 268
hazardous waste disposal and, 27
industrial chemicals and, 197
limitations of, 135–143
selection of sites, 88
sources of wastes, 137

Technology, 249–270
aerial/space detection of pollution, 253–257
agriculture, 261–264
biotechnology, 264–267
ecological impact and, 246
greenhouse gases and, 257–261
hazardous waste disposal and, 131–132
incineration and, 190
international cooperation and, 238
laser detection of groundwater pollution, 249–252
nuclear power production, 229
pollutant detection by, 253–257
pollutant removal and, 24
social engineering and, 267–270
Testing problems
ecological resources and, 86–89
Love Canal, New York, 55–61
uncertainty in, 89–91
See also Hazardous level assessment
Thatcher, Margaret, 236
Third world: *see* Developing nations
Thoreau, Henry David, 271
3M Company, 189–193, 198
Three Mile Island, 97, 110–111, 114, 149, 230
Timber industry, chemical runoff and, 175
Times Beach, 41–43, 91
Tin, water pollution by, 24
Tocqueville, Alexis de, 213
Toluene, characteristics of, 32
Toxic Catastrophe Prevention Act (New Jersey), 171
Toxic substances
categorization for regulation of, 48–52, 61–62
dangers of, v
definition of, 8–9
ecological resources and, 86–89
groundwater contamination by, 160
hazards of, ix–x
international cooperation and, 237
iegislation and, 8
measurement of, 38
mixtures of chemicals, risk assessment, 84–86
potential quantity of, 35
state agencies and, 176–180
transportation accidents and, 206–208
Toxic Substances Control Act of 1976, 1, 28, 238
environmental awakening and, 10–16
politics and, 16–18
premanufacturing notification and, 13, 22
Trade associations, public relations and, 204–205
Trade secrets, protection of, 14
Transportation accidents, dangers of, 206–208
Trichloroethylene, regulation of, 46

Underground storage tanks, hazardous waste disposal, 147–149
Union Carbide, 97–98
Union of Soviet Socialist Republics
Chernobyl disaster, 230–231
cooperation with, 241
national security and, 245
United Nations, 246
environmentalism and, 5, 240

United Nations (*cont.*)
greenhouse gases and, 224, 226
international cooperation and, 241–242
U.S. Consumer Product Safety Commission, 214
U.S. Department of Agriculture, 267
U.S. Department of Energy
Nevada Test Site, 113–114, 151
nuclear waste disposal and, 152–154
nuclear weapons plants and, 156–158
openness and, 126
public trust and, 113
U.S. Department of State, policies on
CFC aerosols, 17
pollutants from Mexico, 241
U.S. Environmental Protection Agency (EPA)
air pollution and, 25–26
air pollution standard setting and, 44–48
budget cuts and, 21–22
burden of proof and, 11–12, 13
community outreach programs of, 113
cost/benefit analysis and, 11
creation of, 5
data release by, 118–122
dioxin and, 41, 43
employee competence and, 133–134
environmental organizations and, 23
groundwater pollution and, 249–252
hazardous level assessment and, 8
hazardous waste disposal analytical problems, 52–55
hazardous waste surveys by, 129–135
hexachlorobenzene and, 35–40
industry waste minimization efforts and, 200–203
industry openness requirements, 204
industry relations with, 197–199
judiciary and, 94–98
Love Canal, New York, and, 55–61
master list of chemicals and, 19–20
mediation and, 125
ozone depletion and, 213–221
PCBs, 72–76
pesticides and, 12–13
policy, promotion of, 103
U.S. Environmental Protection Agency (*cont.*)
politics and, 277–278. *See also* Politics
pollutant removal technology and, 24
pollution detection by technology, 253–257
premanufacturing notification concept, 13
public health and economics, 37
public relations and, 125–128
public trust and, 115–118, 277
research budget of, 268–269
responsibility of, vi
risk analysis by quantification and, 76–84
risk communication, 125–128, 204
risk communication training, 106
shortcomings of, 22–23
state agencies and, 170
Superfund and, 27, 135–143
testing efforts of, 21
toxic chemical categorization and, 49–51
vinyl chloride and, 65–72
water pollution and, 23–25
U.S. Food and Drug Administration
chlorofluorocarbons and, 214
hexachlorobenzene and, 35–40

Vietnam War, 6, 40
Vinyl chloride
air pollution standards, 46
emissions of, 26
media and, 104–105
cancer and, 65–72
Volpe, John, 5

Waste disposal. *See* Hazardous waste disposal
Water pollution
aerial/space detection of, 254–256
carcinogens and, 25
data release, 119
dioxin, 43
ecological resources, 87–88
environmental awakening and, 5
EPA and, 23–24

Water pollution (*cont.*)
international cooperation and, 237–238, 241
legislation and, 8, 9
Natural Resources Defense Council and, 23
persistence of, 272
state agencies and, 173–176
withholding data and, 119
Water quality, dams and, 2

World Bank, 240, 242, 246
World Health Organization, hexachlorobenzene and, 37–38
World Meterological Organization, 224
Wright, Frank Lloyd, 249

X-ray fluorescence, 53

Yucca Mountain, 151–156